CONDUCTIVE POLYMERS

POLYMER SCIENCE AND TECHNOLOGY

Recent volumes in the series:

A Continuation Order Plan is available for this series. A continuation order will bring delivery of each new volume immediately upon publication. Volumes are billed only upon actual shipment. For further information please contact the publisher.

CONDUCTIVE POLYMERS

Edited by
Raymond B. Seymour
University of Southern Mississippi
Hattiesburg, Mississippi

PLENUM PRESS • NEW YORK AND LONDON

Library of Congress Cataloging in Publication Data

Main entry under title:

Conductive polymers.

   (Polymer science and technology ; v. 15)
   Proceedings of a symposium sponsored by the American Chemical Society
Division of Organic Coatings and Plastics Chemistry, held Aug. 26 – 27, 1980, at
the Second Chemical Congress of the North American Continent in Las Vegas, Nev.
   Bibliography:
   Includes index.
   1. Polymers and polymerization—Electric properties—Congresses. I. American
Chemical Society. Division of Organic Coatings and Plastics Chemistry. II.
Chemical Congress of the North American Continent (2nd : 1980 : Las Vegas,
Nev.) III. Series.
QD381.9.E38C66   620.1'9204297   81-12025

ISBN-13: 978-1-4613-3311-1   e-ISBN-13: 978-1-4613-3309-8   AACR2

DOI: 10.1007/978-1-4613-3309-8

Proceedings of a symposium on Conductive Polymers sponsored
by the American Chemical Society Division of Organic Coatings
and Plastics Chemistry, held August 26-27, 1980, at the
Second Chemical Congress of the North American Continent
in Las Vegas, Nevada

© 1981 Plenum Press, New York
Softcover reprint of the hardcover 1st edition 1981

A Division of Plenum Publishing Corporation
233 Spring Street, New York, N. Y. 10013

PREFACE

Those who recognize that our modern life style is dependent, to
a large extent, on the use of organic polymers as thermal and electrical insulators, may be surprised to learn that specific plastics
may also be used as conductors of electricity. In addition to demonstrating the versatility of polymers, this use as conductors will
lead to developments which were not possible with other available
materials of construction.

This is a new field which is growing rapidly because of intensive
research and developmental efforts by many different industrial,
governmental and university investigators. Many of these researchers
reported advances in this art at a symposium on conductive polymers
sponsored by the American Chemical Society's Division of Organic
Coatings and Plastics Chemistry held at the Second Chemical Congress
of the North American Continent at Las Vegas, in August 1980. The
proceedings of this timely symposium are presented in this book.

The editor wishes to take this opportunity to express his grati-
tude to the authors who contributed to this book and to the ACS
Organic Coatings and Plastics Division for sponsoring this effort.

Raymond B. Seymour
Department of Polymer Science
University of Southern Mississippi
Hattiesburg, MS 39401

CONTENTS

NEW HORIZONS IN CONDUCTIVE POLYMERS

Raymond B. Seymour

Department of Polymer Science
University of Southern Mississippi
Hattiesburg, MS 39401

One of the distinguishing features among materials, classified by the ancients as animal, vegetable or mineral, was the ability to serve as conductors of heat. The animal and vegetable classes were covalently-bonded organic compounds which were characterized by their inability to conduct heat. After the discovery of electricity, these compounds were also found to be nonconductors of electrical current, and hence they could be used as insulators for both heat and electricity.

In contrast, the minerals, or at least the metals in this classification, were excellent conductors of heat and electricity. Much of our modern industrial progress has been based on the use of metals, such as copper, aluminum, silver and mercury, as conductors and of organic polymers, such as polyethylene and polystyrene, as insulators.

That nonmetallic materials could store electrostatic electricity was recognized by Thales of Miletus at least 2500 years ago. He demonstrated that amber became charged by rubbing and, of course, the term electricity was coined from the Greek word elektron, meaning amber. Gilbert expanded this concept and charged a series of nonconductors by this rubbing technique or electrification in 1600 A.D.

Otto van Gueriche developed a machine for the production of static electricity by rubbing a ball of sulfur. Since most classical organic polymers, i.e., elastomers or rubbers, fibers and plastics like amber, are also nonconductors, they store electrostatic charges. This phenomenon accounts for the attraction of dust by plastic articles such as blow-molded bottles, the accumulation of electrical charges in carpets in dry weather and radio-frequency and electromagnetic interference (RFI, EMI) of plastic structures.

The tendency for polymers to store electrostatic charges was overcome in the past by the addition of carbon black or acetylene black to natural rubber and metallic powders to plastics and by the blending of metallic fibers or metal-coated fibers with organic fibers.[1] Polymeric composites that were fair conductors of heat and electricity were produced by adding larger amounts of conductive materials to elastomers and plastics.[2-5]

More recent investigations on the effect of metals and salt additives on conductivity by Taylor, short conductive fibers by Bigg, metallized glass fibers (metalloplastics) by Davenport and metal alloy high aspect ratio flakes by Simon are described in chapters in this book.

Macrocyclic tetrazannulene complexes with metallic ions, such as Ni^{2+} and Pd^{2+}, are also good conductors.[6] The synthesis and properties of these organic conductors are described in a subsequent chapter by Hatfield.

Cationic and anionic charged microgel polymers, group IV B metallocene polyoximes and metallic salts of ionomers[7] also provide conductivity in polymeric systems. The first two systems are described in chapters by Upson and Carraher.

Polymers produced by the quenching of sulfur nitride vapor are also conductive. These polymers, which have a ceiling temperature of 145°C, vary in color from dark blue to golden depending on the method used for their preparation.[8,9]

Onnes discovered that mercury was a superconductor some 70 years ago, and more recently Jerome and Beckgaard have reported the superconductivity of ditetramethyltetrathioselenofulvalene hexafluorophosphate $(TMTSF)_2PF_6$. This organic compound has a zero resistance when subjected to high pressure at 0.9K.[10]

It was observed in 1950 that β-carotene, which has an orange color related to the presence of 11 conjugated bonds, exhibits semiconductive and photoconductive properties.[11] It was also noted in the early 1950's and 1960's that polyacetylene[12] and polyphenylacetylene[13] were semiconductors. Newer developments in these polymeric semiconductors are discussed in subsequent chapters.

It had also been observed previously that the rate of thermal dehydrochlorination of polyvinyl chloride could be monitored by noting the formation of colored products and that this discoloration was related to the production of a conjugated polyene structure.[14]

Comparable colored polyenes were also obtained by the thermal dehydration of polyvinyl alcohol, the thermal dehydrocyanation of polyacrylonitrile and by the Ziegler-Natta type polymerization of

1,6-heptadiyne.[15] More information on the electronic structure and
spacial and electronic relationships of extended electron systems
is provided in a chapter by Marks.

While polyacetylene is the most widely investigated conductive
polyene,[16] many other polymers with delocalized π electrons have
been studied. Considerable information on the synthesis and properties of polyacetylenes is provided in chapters by MacDiarmid, Gibson,
Deits, Karasz, Wnek, Chien and co-workers.

Other related conductive polymeric systems are the polyimides,[17]
poly-N-vinylcarbazole,[18] polyphenylene,[19] polypyrrole,[20,21]
N,N-(1,4-phenylene-dimethylidene) bis (3-ethynylaniline),[22] and polyphenylene sulfide. Additional information on the conductivity of
polyimides, polypyrroles and polyphenylene sulfide is provided in
chapters by Taylor, Diaz, Baughman, Chance, Shacklette and Clarke.
Information on the conductivity of the reaction products of polyvinyl ketones with phosphoryl chloride and across the interface
between selenium and polystyrene-polymethyl methacrylate is provided
by Ogawa and Josefowicz.

Polyvinylidene fluoride and polyvinyl fluoride have piezoelectric
properties and polyvinyl chloride and polyacrylonitrile have ferro-
electric properties. Thus, these polymers may be used as trans-
ducers.[23]

The potential usefulness of conductive polymers has been
increased dramatically by the discovery that the electrical conductivity could be increased by several orders of magnitude by "doping,"
i.e., the addition of electron acceptors (p-type) such as iodine or
arsenic pentafluoride[24] and by the addition of electron donors
(n-type), such as sodium or sodium naphthalide.[25] The effect of
this "doping" technique[26] on the conductivity of polyacetylenes,
polyphenylene and polyphenylene sulfide are also discussed in subsequent chapters in this book.

At the present time, the most extensive technological application
of conductive polymers is in the electrophotographic industry. For
example, polymers which are nonconductors in the dark, can conduct
current when exposed to light.[27] Both p- and n-type materials are
being produced by "doping" and specific polymers are being used as
photo- and piezo-conductors.[28] In addition to the conductive potentials for polymers, such as polyacetylene, or polydiacetylene also
has potential applications. It changes color during conformational
transitions to different processes.[29]

MacDiarmid and Heeger, who are co-authors of a chapter in this
book, have already demonstrated that thin sheets of "doped" polyacetylene can convert sunlight to electricity and hence have
potential for use as solar converters as well as for batteries.

Perhaps the most promising prediction of things to come from conductive polymers is summed up in the statement of Nobel laureate John R. Schreiffer who states, "This is the hottest thing in modern physics."

REFERENCES

1. R. H. Norman, Conductive Rubbers and Plastics, Applied Science Publishers Ltd., London (1970).
2. A. R. Blythe, Electrical Properties of Polymers, Cambridge Univ. Press, Cambridge, England (1974).
3. C. A. Klein, Organic Semiconductors, The Macmillan Co., New York (1962).
4. E. Guttman and L. E. Lyons, Organic Semiconductors, John Wiley, New York (1967).
5. N. F. Mott and E. A. Davis, Electronic Processing in Noncrystalline Materials, Oxford Univ. Press, New York (1979).
6. W. E. Hatfield, Molecular Metals, NATO Conference Series, Series VI: Materials Science, Vol. 1, Plenum, New York (1979).
7. R. B. Seymour, F. W. Hayward and I. Brannum, Ind. Eng. Chem. 41, 1479, 1482 (1949).
8. A. G. MacDiarmid and A. J. Heegar, Polymer Preprints 18 (1), 854 (1977).
9. M. Goehring and D. Voight, Naturwissenschaften 40, 482 (1953).
10. K. Bechgaard and D. Jerome, Ind. Res. Dev. (6), 88 (1980).
11. J. Rose and F. Stratham, J. Chem. Soc. 69 (1950).
12. M. Hatano, J. Polym. Sci. 51, 526 (1961).
13. E. Menafee and Y. Pao, J. Chem. Phys. 36, 3477 (1962).
14. C. S. Marvel, J. H. Sample, and M. F. Ray, J. Am. Chem. Soc. 61, 3241 (1939).
15. J. K. Stille and D. A. Frey, J. Am. Chem. Soc. 83, 1697 (1961).
16. I. Ito, H. Shirakawa and S. Ikeda, J. Polym. Sci. 12, 11 (1974).
17. V. C. Carver, T. A. Fursch, L. T. Taylor and A. K. St. Clair, Organic Coatings and Plastics Chemistry 41, 150 (1979).
18. J. Mort, G. Pfister, S. Grammatica and D. J. Sandman, Solid State Comm. 18, 693 (1976).
19. R. H. Baughman, D. M. Ivory and G. G. Miller, Organic Coatings and Plastics Chemistry 41, 139 (1979).
20. A. F. Diaz, K. K. Kanazawa and C. P. Cardini, J. Chem. Soc. Chem. Comm., 655 (1979).
21. K. K. Kanazawa, A. F. Diaz, R. H. Geiss, W. I. Gill, J. F. Kwak, L. A. Logan, J. F. Rabalt and G. B. Street, J. Chem. Soc. Chem. Comm., 854 (1979).
22. T. H. Walton, Organic Coatings and Plastics Chemistry 42, 595 (1980).
23. M. G. Broadhurst, S. Edelman and G. T. Davis, Organic Coatings and Plastics Chemistry 42, 241 (1980).
24. T. Ito, H. Shirakawa and S. Ikeda, J. Plym. Sci., Polym. Chem. Ed. 12, 11 (1974).

25. T. C. Clarke, R. H. Geiss, J. F. Kwak and G. B. Street, J. Chem. Soc. Chem. Comm. 489 (1978).
26. J. Mott, Science (208), 819 (1980).
27. J. Weigl, Angew Chem. Int. Ed. Engl. 16, 374 (1977).
28. T. Tani, P. M. Grant, W. D. Gill, G. B. Street and T. C. Clarke, Solid State Comm. 33, 499 (1980).
29. J. H. Krieger, Chem. Eng. News 58, 25 (1980).

SYNTHESIS AND CHARACTERIZATION OF CONDUCTIVE PALLADIUM

CONTAINING POLYIMIDE FILMS

T.L.Wohlford, J.Schaff, L.T.Taylor, A.K. St. Clair
T.A.Furtsch and E.Khor
Chemistry Department   NASA Langley Research
Virginia Polytechnic Institute   Center
and State University   Hampton, VA 23665
Blacksburg, VA 24061

INTRODUCTION

The doping of neutral polymers with dissolved metal salts or metal complexes for the purpose of imparting electrical conductivity to the polymer has received little attention. In the few instances where this technique has been applied enhancement of conductivity has been marginal. Angelo[1] initially reported in a patent the synthetic procedure for the addition of metal ions to numerous types of polyimides. Few properties of the films which he cast are available from the patent. However, the electrical properties at room temperature of a film derived from 4,4'-diamino-diphenylmethane, pyromellitic dianhydride and bis(acetylacetonato)-copper(II) were given as follows: percent copper, 3.0%; dielectric constant, 3.6%; dissipation factor, 0.004-0.01; volume resistivity, 8×10^{12} ohm-cm. The measured resistivity represented a 5-6 orders of magnitude enhancement over the polymer alone. Further patents or published work in this area are not available.[2]

Superior antistatic properties have been reported[3] for a newly available soluble polyimide (DAPI-Polyimide) film to which had been added a low level of either $LiNO_3$ or $LiCl$. Film physical properties and smoothness remained unchanged except that electrical resistance was sharply lowered. Conductivity was increased approximately 2000% over the standard unfilled polyimide. These lithium doped films were observed to be slightly hygroscopic which may account for the reduced resistivity.

A series of metal salts having poly(alkylbenzimidazoles) as the parent ligand have been synthesized.[4] Values for the electrical resistivity did not significantly change upon doping the

7

polymer with either $CrCl_3 \cdot 6H_2O$, $CoCl_2 \cdot 6H_2O$ or $Ni(C_2H_3O_2)_2 \cdot 4H_2O$. In contrast treatment of the polymer with HCl thereby forming the polymer conjugate acid changed the volume resistivity from 10^{13} ohm-cm to 10^6 ohm-cm. X-ray photoelectron spectra of the core levels of nitrogen and metal suggested the formation of polybenzimidazole-metal complexes.

The surface electrical conductivity of films of poly(vinyl-alcohol)-Cu(II) and poly(acrylamide)-Cu(II) complexes has been shown[5,6] to increase by spreading an acetone solution of iodine over the film surface. After iodine treatment a whitish substance identified as γ-CuI appears on the film surface. The strongly adhering γ-CuI is believed to be responsible for the enhanced electrical conductivity.

Glassy polymers have been produced from[7] solutions of $Ca(NCS)_2$ and the polymer derived from Bisphenol A and epichlorohydrin. Increased electrical conductivity was found to result from salt incorporated into the polymer. Surface and volume resistivities for the polymer alone were 1×10^{15} ohm and 5×10^{13} ohm-cm; whereas, for "phenoxy" polymer with 13% $Ca(NCS)_2$ by weight the values were 1×10^{10} ohm and 4×10^{10} ohm-cm. The decrease in resistivity was attributed to the high equilibrium water content accompanying calcium ion addition.

In our laboratory the doping of polyimides with metallic species has been demonstrated to (1) increase the softening temperature, (2) increase high temperature adhesive properties, and (3) change the polymer decomposition temperature.[8] We wish to report here the synthesis and characterization of a series of palladium-filled polyimide films many of which exhibit significantly lowered surface and volume resistivities. A preliminary communication on one of these polyimide films has recently appeared.[9]

EXPERIMENTAL

Materials – Pyromellitic dianhydride (PMDA) and 3,3',4,4'-benzophenone tetracarboxylic acid dianhydride (BTDA) were obtained from commerical sources and purified by sublimation at 215°C at less than 1 torr, melting point 497°K and 558°K respectively. Oxydiani-line (ODA) and 3,3'-diaminobenzophenone (DABP) were obtained from commercial sources and purified by recrystallization.[10] 3,3'-Diamino-diphenylcarbinol (DADPC) was prepared by reduction of DABP. Equimolar quantities of 3,3'-DABP and sodium borohydride were dis-solved in diglyme. Small amounts of ethanol and water were added to serve as proton sources for the reaction. The DABP solution was allowed to stir for approximately 20 hours. 6M HCl was added slowly and the pH was adjusted to 6-7. The solution was then extracted with brine and THF. The organic layers were removed by flash evaporation and a waxy yellow compound was obtained. The wax was dissolved in water and recrystallized. The resulting

3,3'-DADPC was very pure with a melting point of 125°C as determined by differential thermal analysis at 5°C/min. (lit. value is 125°C). An infrared spectrum of the product showed a strong -OH peak at 3550 cm^{-1}. The monomer was thus considered to be polymer grade. N,N-Dimethylacetamide (DMAC) was obtained from Burdick and Jackson and used as received. The solvent was reagent grade, distilled in glass and packed under N_2. Lithium tetrachloropalladate(II) (Li_2PdCl_4) was prepared by a previously reported procedure.[11] Dichlorobis (dimethylsulfide)palladium(II), $(Pd(S(CH_3)_2)_2Cl_2)$, was synthesized by slightly modifying a reported[12] procedure for the analogous platinum complex. All other chemicals were of reagent grade or equivalent and were obtained from commercial sources.

Polymerization - Polymerizations were carried-out in solutions containing 20% solids (w/w) by adding diamine (0.004 mole) and DMAC to a flask flushed with dry nitrogen. The appropriate dianhydride (0.004 mole) was then added as a solid in a single portion and the solution was stirred at room temperature for 30-36 hours. At this point the appropriate metal complex (0.001 mole) was added directly or added as a solution in a minimum amount of DMAC to the polyamic acid-DMAC solution. In most cases polymerization conducted in situ with the appropriate metal complex gave similar results. Polymer-metal complex solutions were refrigerated until cast.

Preparation of Films - Polyamic acid-metal complex solutions were poured onto soda-lime glass plates. Solutions were spread using a doctor blade with a 8-16 mil gap to ensure a final film thickness of approximately 1 mil. Films of the polyamic acid-metal complex were dried in static air at 60°C for two hours. Imidization was thermally achieved by heating in a forced air oven 1 hour each at 100°, 200° and 300°C. A specially constructed aluminum box was employed for imidization in an anerobic environment. Polyimide-metal complex films were removed from the glass plate by soaking several hours in warm water.

Characterization - Thermal mechanical analyses (TMA) were performed on films in static air at a 5°C/min temperature program on an E.I. DuPont Model 990 Thermomechanical Analyzer. Thermogravimetric analyses (TGA) were obtained on films at 2°C/min. in static air. Surface and volume resistivities were measured following the standard ASTM method of test for electrical resistance of insulating materials (D 257-66) employing a Keithley voltage supply, electrometer and four-point probe. X-ray photoelectron spectra (XPS) were obtained on a DuPont 650B spectrometer equipped with a Mg anode (Mg K_α=1253.6ev) target. The binding energies of all electrons were measured relative to the instrumental background carbon ($1s_{\frac{1}{2}}$) photopeaks taken to have a value of 284.0 ev. Scanning electron microscopy was performed with an AMR 900 Scanning Electron Microscope utilizing an EDAX 707A X-Ray Analyzer. Graphite was vapor deposited onto the film surface prior to analysis.

RESULTS AND DISCUSSION

Palladium-filled polyimide films have been prepared using the following dianhydride-diamine pairs: BTDA + ODA, BTDA + DABP, BTDA + DADPC, PMDA + ODA, PMDA + DABP and PMDA + DADPC (Structures I-V). A number of palladium additives were screened many of which proved unacceptable because of insufficient solubility in DMAC or in the polymer-DMAC solution. While good quality films could be produced with slightly soluble $PdCl_2$ and Na_2PdCl_4 and even dispersed palladium metal, only minor modifications could be realized in polymer properties. In many of these cases non-uniform features in the film were apparent. Best results todate have been obtained with Li_2PdCl_4 and $Pd(S(CH_3)_2)_2Cl_2$ as additives. Reasonably good quality, dark red-brown films have been fabricated for the six monomer pairs noted below.

The synthetic procedure involved formation of the polyamic acid, VI, in DMAC, intimate mixing of the palladium complex and polyamic acid, and thermal imidization in air to the palladium-filled polyimide film. An alternate _in situ_ method whereby polymerization to the polyamic acid was performed in the presence of

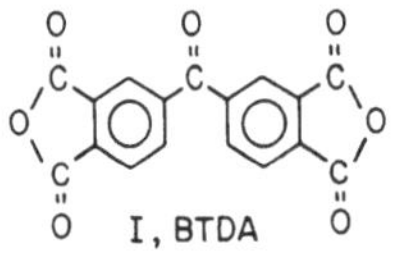

I, BTDA

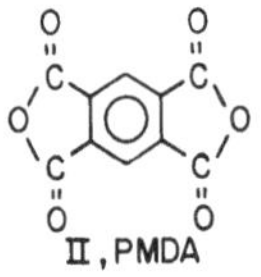

II, PMDA

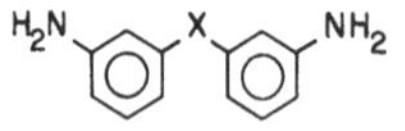

III, DABP (X = C=O)
IV, DADPC (X = CH-OH)

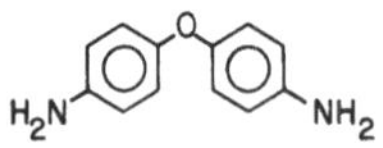

V, ODA

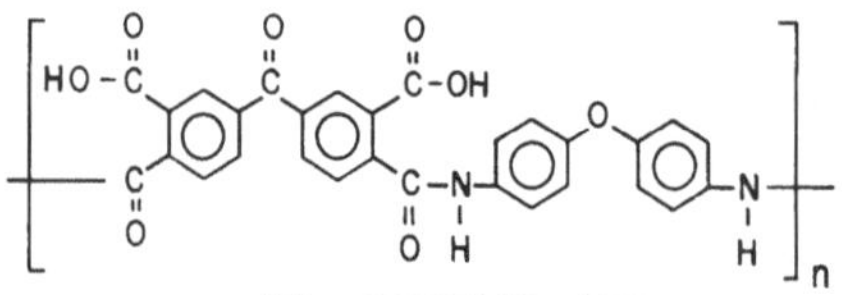

VI , POLYAMIC ACID

the palladium complex also proved satisfactory. The films, as removed from the glass plate, many times possessed noticeably different surfaces depending upon whether the film had been exposed to the glass or to the air during the imidization procedure. This difference was very noticeable for the two $Pd(S(CH_3)_2)_2Cl_2$ containing films BTDA + ODA and BTDA + DABP. While the glass side had a dark red-brown appearance, the air-side possessed a definite silvery, metallic appearance. Re-heating of these films after removal from the glass plate caused the glass-side to also metallize. The metallic looking surfaces could not be produced with either PMDA, DADPC or Li_2PdCl_4 containing films. The presence of oxygen during the imidization process appears crucial, since BTDA + ODA and BTDA + DABP doped with $Pd(S(CH_3)_2)_2Cl_2$ do not give rise to metallic surfaces when cured in either a dry N_2, Ar, N_2/H_2 or moist Ar atmosphere.

Regardless of the additive, monomer pair or curing atmosphere, the films are essentially static free, moisture free and relatively thermally stable. The PMDA derived polymers yielded more brittle films which is not surprising due to their rigid chemical structure and high modulus. However, the technique of casting thinner films produced greater film flexibility.

Thermomechanical (TMA) and thermogravimetric (TGA) data have been obtained on most of the films. Table I lists the apparent polymer decomposition temperatures from TGA for polymer alone and palladium-filled films. Thermal stability is reduced by approximately 25% for Li_2PdCl_4 films and approximately 40% for $Pd(S(CH_3)_2)_2Cl_2$ films except for DADPC films where the two complexes yielded similar data. Data on films cured in nitrogen were about the same within experimental error. It is unclear why the sulfur containing additive gives rise to a less thermally stable film since the monomer-additive mole ratio was always 4:1. The effect of the two palladium complexes on the apparent glass transition temperature from TMA was not as significant except again for the BTDA + DADPC + $Pd(S(CH_3)_2)_2Cl_2$ film which showed an 84°C rise. In general, the AGT was unchanged or increased. The Li_2PdCl_4 was least deleterious in modifying the highly desirable thermal and mechanical properties of the BTDA + ODA and BTDA + DABP polyimides. Naively one might attribute these results to the apparently higher experimentally determined palladium content in the $Pd(S(CH_3)_2)_2Cl_2$ films. Since the two palladium complexes are of comparable molecular weight and all filled polyimides are prepared employing a 4:1 mole ratio the % Pd should not appreciably differ. Calculated % Pd for the filled BTDA polyimides range from 4.5-5.2%. The Li_2PdCl_4 films yield %Pd in the 4-5% range. This is reasonable since we have only subtracted the water of imidization in our %Pd calculation. Elemental analyses suggest that a significant amount of Li and Cl has been lost during the curing process which could account for the

TABLE I

THERMOGRAVIMETRIC AND THERMOMECHANICAL
DATA FOR PALLADIUM-FILLED POLYIMIDES

Film	PDT[a]	AGT[b]	%Pd
BTDA + ODA	540	286	-
BTDA + ODA + Li_2PdCl_4	411(392)[c]	336(326)	4.8(4.2)
BTDA + ODA + $Pd(S(CH_3)_2)_2Cl_2$	317(351)	317[d](357)[d]	7.9(5.0)
BTDA + DABP	570	257	-
BTDA + DABP + Li_2PdCl_4	410(403)	278(252)	5.0(4.89)
BTDA + DABP + $Pd(S(CH_3)_2)_2Cl_2$	320(372)	253(258)	6.3(5.00)
BTDA + DADPC	547	307	-
BTDA + DADPC + Li_2PdCl_4	375(432)	299(270)	4.54(4.57)
BTDA + DADPC + $Pd(S(CH_3)_2)_2Cl_2$	377(350)	391[d](361)[d]	4.96(4.81)
PMDA + ODA	580	405	-
PMDA + ODA + Li_2PdCl_4	-	-	-
PMDA + ODA + $Pd(S(CH_3)_2)_2Cl_2$	340(360)	401[d](394)[d]	5.86(6.41)
PMDA + DABP	-	321	-
PMDA + DABP + Li_2PdCl_4	-	-	-
PMDA + DABP + $Pd(S(CH_3)_2)_2Cl_2$	- (360)	- (390)[d]	- (5.99)
PMDA + DADPC	580	340	-
PMDA + DADPC + Li_2PdCl_4	393(396)	371[d](346)	- (5.79)
PMDA + DADPC + $Pd(S(CH_3)_2)_2Cl_2$	353(355)	393[d](385)	5.30(5.24)

[a]Polymer decomposition temperature (°C)

[b]Apparent glass transition temperature (°C)

[c]Numbers in parenthesis correspond to data on polymers cured in nitrogen

[d]With decomposition

lower than calculated values. The situation regarding $Pd(S(CH_3)_2)_2Cl_2$ is difficult to explain since many of these films show %Pd > 5.2%. A more heterogeneous film is one obvious rationalization for these data; while significant polymer decomposition may have occurred during curing since the PDT for these polymers are just above 300°C.

The primary purpose for this study was to ascertain if palladium-filled polyimides exhibited lower resistivities than unfilled polyimides. Table II outlines these results. For BTDA derived films, four different combinations of dianhydride, diamine and additive yielded dramatically lowered resistivities. With PMDA only one combination was successful even though all good quality films were prepared in an identical manner. Surprisingly Li_2PdCl_4 gave a lowered resistivity value with BTDA + DADPC; while, $Pd(S(CH_3)_2)_2Cl_2$ with the same monomer pair exhibited a polymer-alone resistivity value. The results with BTDA + DABP, however, were reversed. Both additives with BTDA + ODA gave conductive films. The metallic surface on one side displayed by the two conductive $Pd(S(CH_3)_2)_2Cl_2$ films, no doubt, reduces resistivity (<10^5 ohm-cm). An exact measurement could not be made with our resistivity measuring device. The sulfur-containing additive with BTDA + DADPC and all PMDA films, on the other hand, gave no metallic surface and no resistivity lowering. The metallic surface is apparently not a necessity for improving conductivity, since some Li_2PdCl_4 films are conductive and none show a metallic surface. Although as Table II attests, resistivities are higher for Li_2PdCl_4.

The results on curing the films in a non-oxygenated atmosphere are equally interesting, Table III. No metallic surfaces are produced with $Pd(S(CH_3)_2)_2Cl_2$ as an additive and no resistivity lowering is observed. The presence of dioxygen seems crucial in this regard. Moist argon and forming gas (N_2/H_2) give the same unchanged results. A nitrogen curing atmosphere, however, does not change the resistivity results appreciably from the air-cured, conductive Li_2PdCl_4 films. Compare Tables II and III. It is significant that in each case, with Li_2PdCl_4, the resistivity values are always one to three orders of magnitude higher for nitrogen cured films. We, therefore, believe that dioxygen is crucial for production of the most conductive films, however, the chemistry of the two additives during the imidization process may be subtly different.

X-ray photoelectron spectroscopy (XPS) has proven valuable in studying some of these palladium-filled polyimides. Measured XPS binding energies (Pd $3d_{5/2, 3/2}$) indicate that an appreciable amount of palladium has been reduced to the elemental state in most films. In other words, during the imidization process reduction of palladium has occurred. In general, those films which have lowered resistivities exhibit the most surface reduced palladium with the exception of BTDA + DADPC + Li_2PdCl_4. Differences between air and glass for the conductive films are very apparent here again,

TABLE II

SURFACE[a] AND VOLUME[b,c] RESISTIVITIES[d]
OF CONDUCTIVE PALLADIUM-FILLED
POLYIMIDES PREPARED IN AIR

Polymer Formulation	Metal Complex	ODA	DABP	DADPC
BTDA	Li_2PdCl_4	9.5×10^5 ohm 2.0×10^6 ohm-cm	NC[e]	1.3×10^7 ohm 1.0×10^7 ohm-cm
	$Pd(S(CH_3)_2)_2Cl_2$	$<10^5$ ohm $<10^5$ ohm-cm	$<10^5$ ohm $<10^5$ ohm-cm	NC[e]
PMDA	Li_2PdCl_4	NC[e]	NT[f]	8.8×10^9 ohm 3.8×10^{11} ohm-cm
	$Pd(S(CH_3)_2)_2Cl_2$	NC[e]	NT[f]	NC[e]

[a] ohm,   [b] ohm-cm,   [c] Polymer alone surface and volume resistivities are 10^{17} ohm and 10^{17} ohm-cm.

[d] Resistivity values are quoted for the best quality films. Replicate films do not differ by more than one order of magnitude.

[e] NC – No change in resistivity relative to polymer alone.

[f] NT – Not testable because of poor film quality.

TABLE III

SURFACE[a] AND VOLUME[b,c] RESISTIVITIES[d]
OF CONDUCTIVE PALLADIUM-FILLED
POLYIMIDES PREPARED IN A NITROGEN ATMOSPHERE

Polymer Formulation	Metal Complex	ODA	DABP	DADPC
BTDA	Li_2PdCl_4	5.1×10^7 ohm 8.9×10^7 ohm-cm	NC^e	2.1×10^{10} ohm 1.4×10^{11} ohm-cm
	$Pd(S(CH_3)_2)_2Cl_2$	NC^e	NC^e	NC^e
PMDA	Li_2PdCl_4	NC^e	NT^f	1.1×10^{13} ohm 2.8×10^{14} ohm-cm
	$Pd(S(CH_3)_2)_2Cl_2$	NC^e	NT^f	NC^e

[a] ohm,   [b] ohm-cm,   [c] Polymer alone surface and volume resistivities are 10^{17} ohm and 10^{17} ohm-cm.

[d] Resistivity values are quoted for the best quality films. Replicate films do not differ by more than one order of magnitude.

[e] NC – No change in resistivity relative to polymer alone.

[f] NT – Not testable because of poor film quality.

since, more reduced palladium usually accompanies the air side.
See $Pd_{(air)}/Pd_{(glass)}$ counts in Table IV. This is best demonstrated with the BTDA + ODA + $Pd(S(CH_3)_2)_2Cl_2$ film, Figure 1, where
essentially all surface palladium is reduced to the metallic state.

XPS for non-conductive films differs in several respects from
the above. Figure 2 illustrates the Pd $3d_{5/2, 3/2}$ photopeaks
observed in the BTDA + DABP + Li_2PdCl_4 film. First of all, the
palladium signals are realtively weak and the measured binding
energies fall between Pd metal and $PdCl_2$. Secondly, a pair of
photopeaks are observed at approximately 10 eV higher energy (349.8
and 346.1 eV). The relative intensity of the two pair of peaks
has been found to vary depending upon what place on the film was
being sampled. Conductive films do not show this higher energy
pair of photopeaks. We attribute this phenomenon to differential
charging of the surface palladium. In other words, some islands
of palladium are more insulated than other islands by the non-
conductive polyimide and cannot dissipate the photo-charge produced
by the x-ray beam. High resistivity films, therefore, appear to
be highly heterogeneous insofar as palladium is concerned. If
palladium is the charge carrier as we expect, an unequal distribution of palladium can result in an interruption of the charge
transfer mechanism. The addition of higher amounts of palladium
to these films does not produce the desired results. After curing
for three hours at 300°C the XPS spectrum, Figure 3, clearly shows
evidence for Pd(II) and Pd(0) on the surface for 4:2 monomer;
additive ratios. Longer heating times result in more complete
palladium reduction but with significantly more polymer thermal
degradation.

Scanning electron microscopy (SEM) and energy dispersive
x-ray analysis (EDAX) on select films have shown rather interesting
features. EDAX confirms again the greater amount of palladium on
conductive films relative to insulative films. Further, it also
supports the notion that for conductive films the air side has the
greater amount of palladium. At this point mostly conductive films
have been investigated. Even though partly conductive suitable SEM
photographs could be obtained only on films which had previously
been coated with condutive carbon. Figure 4 shows a SEM photo at
500X magnification of BTDA + ODA + Li_2PdCl_4. Variable shaped
particles are observed all over the surface. The "star-like"
deposits as well as the smooth surface show appreciable concentrations of palladium and chlorine with the former in greater concentration. Pictures taken up to 10000X magnification gave no new
information. The glass side of this same film at 1000X gave a
similar rough looking surface (Figure 4) wherein the deposits seemed
less organized and smaller in size. The film is essentially palladium with little detectable chlorine.

The same film cured in nitrogen showed somewhat different

TABLE IV

XPS DATA FOR VARIOUS PALLADIUM CONTAINING BTDA-DERIVED FILMS[a]

Film	Film Side	Counts Pd(air)/Pd(glass)	Binding Energy(eV) Pd 3d $5/2,3/2$	Curing Atmosphere
ODA + Li_2PdCl_4	Air	25	341.2, 336.2	Air
	Glass		340.2, 335.2	
	Air	–	339.4, 334.0	N_2
	Glass		Vwb	
ODA + $Pd(S(CH_3)_2)_2Cl_2$	Air	9.7	339.8, 334.6	Air
	Glass		340.4, 334.7	
	Air	2.0	340.3, 335.0	Argon
	Glass		340.3, 335.0	
DABP + Li_2PdCl_4	Air	5.0	341.8, 336.3	Air
	Glass		339.7, 334.5	
DADPC + Li_2PdCl_4	Air	0.71	343.5, 338.4	Air
	Glass		343.5, 338.2	
	Air	–	Vwb	N_2
	Glass		339.3, 334.3	

[a] Pd 3d $5/2,3/2$ for palladium metal = 340.8 and 335.2 eV and for $PdCl_2$ = 342.6 and 337.3 eV.

[b] VW means very weak, barely detectable signal.

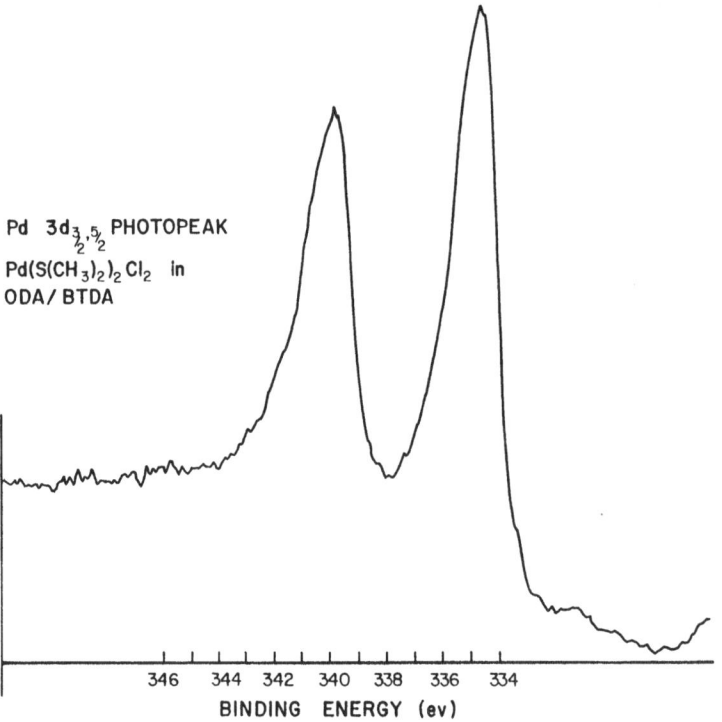

Pd 3d$_{3/2}$,$_{5/2}$ PHOTOPEAK

Pd(S(CH$_3$)$_2$)$_2$Cl$_2$ in
ODA/ BTDA

BINDING ENERGY (ev)

Figure 1

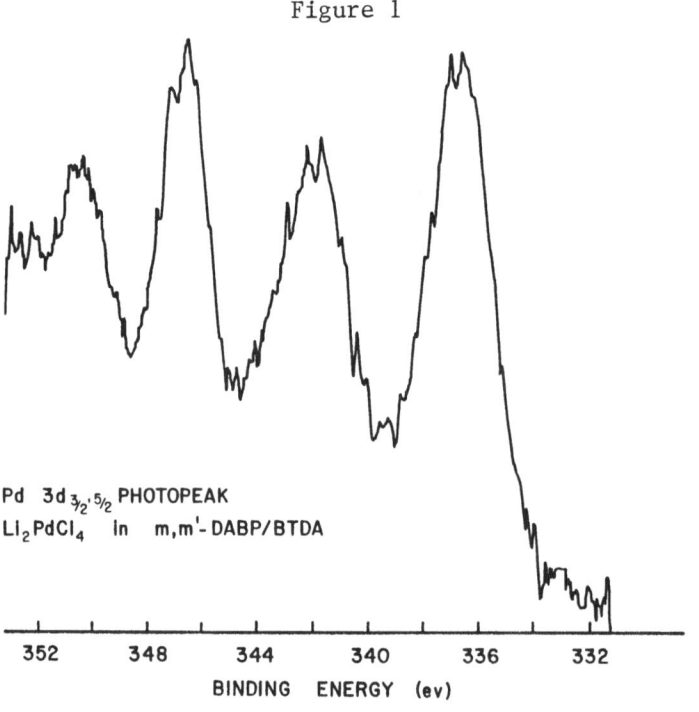

Pd 3d$_{3/2}$,$_{5/2}$ PHOTOPEAK
Li$_2$PdCl$_4$ in m,m'- DABP/BTDA

BINDING ENERGY (ev)

Figure 2

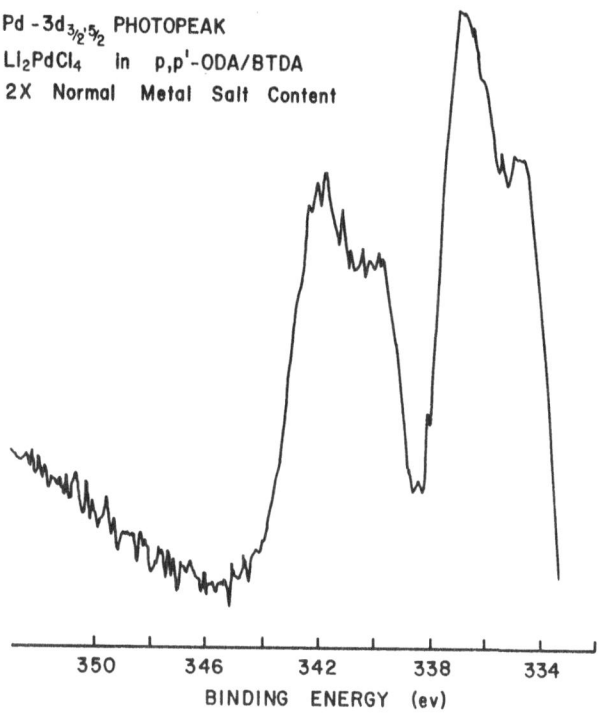

Pd - 3d$_{3/2;5/2}$ PHOTOPEAK
Li$_2$PdCl$_4$ in   p,p'-ODA/BTDA
2X Normal Metal Salt Content

BINDING ENERGY (ev)

Figure 3

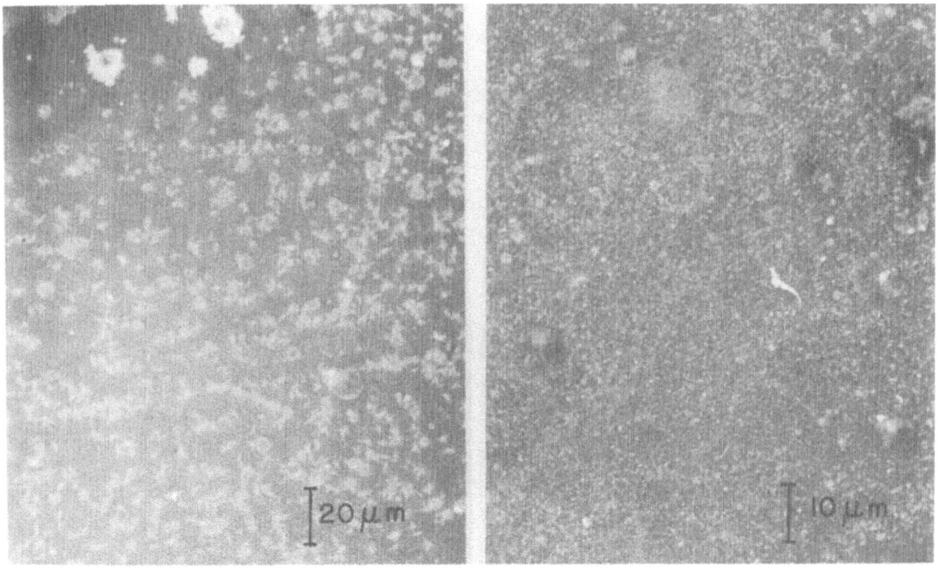

BTDA + ODA + Li$_2$PdCl$_4$ BTDA + ODA + Li$_2$PdCl$_4$
(AIR SIDE - AIR ATMOSPHERE)   (GLASS SIDE - AIR ATMOSPHERE)

Figure 4

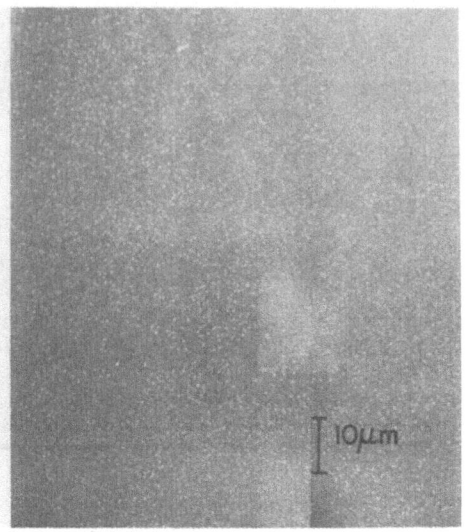

BTDA + DABP + $Pd(S(CH_3)_2)_2Cl_2$
(Air Side - Air Atmosphere)
1000X

BTDA + DABP + $Pd(S(CH_3)_2)_2Cl_2$
(Glass Side - Air Atmosphere)
1000X

Figure 5

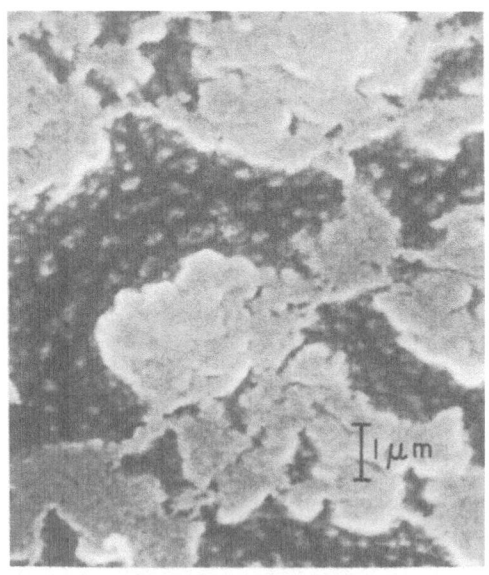

BTDA + DABP + $Pd(S(CH_3)_2)_2Cl_2$
(Air Side - Air Atmosphere)
10000X

Figure 6

features. The SEM photo shows some particles on the surface at 500X magnification which appear to be fewer in number and some of which are quite large. The particles contain predominantly chlorine; whereas, the surface is mainly palladium. The glass side exhibited a relatively low level of palladium compared to the other side and no chlorine was detectable.

The aggregation of particles on the surface is best observed in the SEM of the air side of the BTDA + DABP + $Pd(S(CH_3)_2)_2Cl_2$ film, Figure 5. EDAX shows that the particles as well as the cracks contain palladium and a trace of sulfur. At greater magnification the cracks are revealed to have smaller sized particles, Figure 6. The glass side of this film which is non-metallic contains only finely divided specks of material which at ten-times the magnification become small pores. The specks and general surface contain primarily palladium with some sulfur. This same film, BTDA + DABP + $Pd(S(CH_3)_2)_2Cl_2$, when cured in a nitrogen atmosphere contains no metallic coating on either side and it does not conduct. The SEM photo reveals no distinguishable features or pattern on either the glass or air surface regardless of the magnification. A very small amount of palladium is found on both sides of the film but nothing like a speck or particle could be found.

One, therefore, might conclude that these SEM distinguishable particles are necessary for achieving a lower resistivity film. We cannot at this time be sure of the nature of this deposit. The XPS measurement, no doubt, is sampling both particles and general surface at the same time. While XPS suggests palladium(II) some photopeaks are rather broad. The detection of chlorine or sulfur in the particles could indicate a non-stoichiometric palladium chloride or sulfide mixed with palladium metal. Future work with scanning Auger electron spectroscopy on these films is planned.

Acknowledgement - Informative discussions regarding this work with T. L. St. Clair are gratefully appreciated.

REFERENCES

1. R. J. Angelo and E. I. DuPont de Nemours & Co., "Electrically Conductive Polyimides", U.S. Patent 3 073 785 (1959).
2. R. J. Angelo, Personal Communication.
3. M. N. Sarboluki, "Antistatic Additive for Polyimide Films", NASA Tech. Brief, 3(2), 36 (1978).
4. S. M. Aharoni and A. J. Signorelli, "Electrical Resistivity and ESCA Studies on Neutral Poly(alkyl benzimidazoles), Their Salts, and Complexes", J. Appl. Polym. Sci., 23, 2653 (1979).
5. O. Sumita, A. Fukuda, and E. Kuge, "Coagulation of Polymer-Cu^{+2} Complexes in Polymer Films and Its Application For Producing

Semiconducting CuI Surface Layers", J. Appl. Polym. Sci.,
23, 2279 (1979).

6. F. Higashi, C. Su Cho, H. Kakinoki, and O. Sumita, "Semicon-
    ducting Organic Polymer From Polyacrylamide-Cu^{++} Chelate and
    Iodine", J. Polym. Sci., Polym. Chem. Ed., 15, 2303 (1977).

7. M. J. Hannon and K. F. Wissbrun, "Interaction of Inorganic
    Salts with Polar Polymers. I. Physical Properties of
    Phenoxy-Calcium Thiocyanate Mixtures", J. Polym. Sci., Polym.
    Phys. Ed., 13, 113 (1975).

8. L. T. Taylor, A. K. St. Clair, V. C. Carver and T. A. Furtsch,
    "Incorporation of Metals Ions Into Polyimides", ACS Sym.
    Ser., 121, 71 (1980).

9. J. C. Carver, L. T. Taylor, T. A. Furtsch and A. K. St. Clair,
    "Electrically Conductive Polyimide Films Containing Pal-
    ladium Coordination Complexes", J. Amer. Chem. Soc., 102,
    876 (1980).

10. V. L. Bell, B. L. Stump and H. Gager, "Polyimide Structure -
    Property Relationships. II. Polymers From Isomeric
    Diamines", J. Polym. Sci. Polym. Chem. Ed., 14, 2275 (1976).

11. A. C. Cope and E. C. Friedrich, "Electrophilic Aromatic Sub-
    stitution Reactions by Platinum(II) and Palladium(II)
    Chlorides on N,N-Dimethylbenzylamines", J. Amer. Chem. Soc.,
    90, 911 (1968).

12. G. B. Kauffman and D. O. Cowan, "Cis- and Trans-Dichlorobis
    (Diethyl Sulfide) Platinum(II)", Inorg. Syn., 6, 211 (1960).

CONDUCTIVE POLYMERIC COMPOSITES FROM SHORT CONDUCTIVE FIBERS

Donald M. Bigg and E. Joseph Bradbury

BATTELLE
Columbus Laboratories
505 King Avenue
Columbus, OH 43201

INTRODUCTION

The ability of metals to conduct both electricity and heat is a major reason for their use in many applications. Static buildup on electrical equipment and high-speed machinery is prevented by grounding of metal parts, and heat-generating components are often protected by the dissipation of excess heat through metal conductors. In a similar manner, high-frequency electromagnetic radiation is either prevented from escaping from the equipment or the equipment is shielded from stray radiation by metal shields and covers.

In addition to electrical and thermal conductivity, certain other special properties are frequently required or desired for a particular application. Three such properties are flexibility, low weight, and the ability, through appropriate processing, to form intricate shapes with varying thickness. Die-cast aluminum, zinc, and their alloys are typical materials frequently used to form intricately shaped conductive housings. However, structures made from these metals are not flexible, often weigh more than desired, and recently their costs have risen sharply. For these reasons conductive systems based on nonmetallic materials have been used to provide lightweight, flexible, and moldable parts having good static bleed-off and electromagnetic interference (EMI) shielding properties. A variety of uses for such materials are encountered, ranging from compliant gasketing to rigid housings for business machines. One of the more commonly used systems consists of a molded plastic part coated with a conductive, metallic surface. The coating is obtained by applying a conductive paint or vapor depositing a metal film on the surface of the part. However, such approaches are often costly and long-term performance is not always as good as desired.

The deficiencies of these approaches have led to greater interest
in the conductive composite approach; that is, the addition of a
conducting filler to a nonmetallic matrix.

In this program the properties of composites made with several
different types of conductive fillers were investigated. The types
of fillers investigated included carbon black, metal fibers and
flakes, and metal coated glass fibers. Both thermoplastic and
thermosetting matrices were studied. The principal properties of
interest were electrical resistivity, EMI shielding attentuation,
thermal conductivity, and tensile strength.

MATERIALS

Table 1 lists the polymers used as matrices in this study.
These included three grades of polypropylene, low-density polyethylene, polycarbonate, and a thermosetting polyester. Shell Chemical
Company's 5220 polypropylene is a general purpose injection molding
grade. Hercules' PC072 polypropylene contains a modifier which
enhances its bonding to glass fibers. Armstrong 1110 polypropylene
and USI FN-500 polyethylene are fine powders frequently used in
coating applications. Lexan 101 is a general purpose injection
molding grade of polycarbonate. Aropol 8319 is a low viscosity
polyester resin chosen to reduce bubble entrapment during fiber
incorporation and curing.

Table 2 lists the important characteristics of the fillers used
in this study. A variety of aluminum fillers (fibers and flakes)
were prepared by the melt extraction process.[1] Irregularly shaped
electrolytic copper particles, Vulcan XC-72 carbon black, and 3-mm-
long milled-glass fibers were obtained from the commercial sources
listed in Table 2. The nickel coated glass fibers were obtained by
electroless coating of 6-mm-long glass fibers.[2]

COMPOSITE PREPARATION

Thermoplastic Matrices

All the fillers, except the nickel-coated glass fibers, were
hot roll milled into the polymer. The milling temperature was 120 C
for polyethylene, 200 C for polypropylene and polycarbonate. The
milled composites were stripped from the rolls as a sheet; quenched;
and chopped into 5-mm particles. The gap separation between the
rolls was kept as 3 mm and slow roller speeds were employed in order
to minimize damage to the fibrous fillers. The chopped particles
were compression molded into test specimens of different sizes to
accommodate the various properties investigated. The samples were
molded at the same temperature as used in the milling procedure.

Table 1. Characteristics of Polymers Investigated

Material	Grade	Source	Density	Tm,C	Tg,C
Polyethylene	FN-500	U.S.I.	0.92	110	--
Polypropylene	5220	Shell	0.91	163	--
Polypropylene	PC072	Hercules	0.91	163	--
Polypropylene	1110	Armstrong	0.91	163	--
Polycarbonate	Lexan 101	G.E.	1.20	--	152
Polyester	Aropol 8319	Ashland	1.40	thermoset	--

Table 2. Characteristics of Fillers Investigated

Material	Source	Form	Diameter	Length	Aspect Ratio
Aluminum	BCL	fibers	0.100 mm	1.25 mm	12.5
Aluminum	BCL	fibers	0.127 mm	3.05 mm	24.0
Aluminum	BCL	fibers	0.200 mm	5.60 mm	28.0
Aluminum	BCL	fibers	0.200 mm	7.00 mm	35.0
Aluminum	BCL	flakes	0.076 mm[a]	1.27 x 0.64 mm	16.7[b]
Copper	Alcan	particles	44 μm	--	1.0
Nickel-Glass	BCL	fibers	14 μm	0.50 mm	450/35[d]
Carbon Black	Cabot XC-72	particles	30 nm	--	--[c]
Glass	Corning	fibers	12.5 μm	100 μm	250/8[d]

(a) Thickness
(b) Long dimension divided by thickness
(c) Carbon particles aggregate into chain-like structures of varying aspect ratios
(d) Before processing/after processing

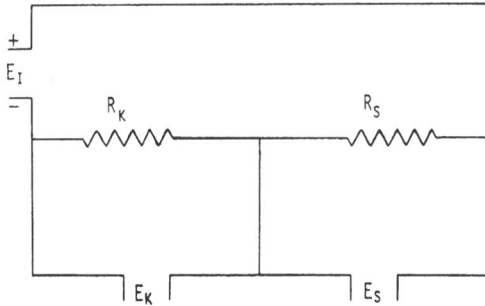

Figure 1. Circuit used to measure the resistivity of highly
           conductive samples.

The nickel-coated glass fibers were dry blended with polycarbonate pellets prior to injection molding into test plaques. The
injection temperature of the melt was 250 C.

Thermosetting Matrix

The fillers were carefully hand mixed into the liquid polyester
resin. Cab-O-Sil was used to increase the viscosity of the less
highly loaded samples in order to prevent filler settling during
cure. Prior to cure the samples were subjected to a vacuum in order
to remove bubbles from the composites. Most of the composites were
cured at 80 C for 1 hour using dicumyl peroxide as the catalyst.
Other samples were cured at various temperatures, using different
concentrations of catalyst, in order to study the influence of conducting fibers on the exotherm and rate of cure.

EXPERIMENTAL

Electrical Measurements

The electrical resistivities of the base polymers and the high-
resistivity composites were measured on a Keithly Instrument Company
6105 resistivity adaptor. The resistivities of the more conductive
composites were measured using the circuit shown in Figure 1. A
standard resistance is connected in series with the sample and a
current is passed through the series circuit. Since the voltage
drop across each resistor is proportional to its resistance, the
resistance of the sample can be determined by comparing the voltage
drop across it with that across a known resistor. Mathematically,
the resistivity of the sample is measured from the following equation:

$$\rho_s = \frac{R_k E_s A_s}{E_k t_s} \quad , \qquad (1)$$

where ρ_s is the sample resistivity and R_k is the resistance of the
known resistor. A_s is the surface contact area of the sample, E_s
is the voltage drop across the sample, E_k is the voltage drop across
the known resistor, and t_s is the thickness of the sample.

The composites were molded between sheets of printed circuit
copper foil: the bonding side has a nodular surface which breaks up
the laminate interface and improves electrical contact. The smooth
exterior of the copper foil is then polished and the part is sandwiched between copper contact plates under 8 kPa pressure. These
contact plates are wired to the circuit shown in Figure 1.

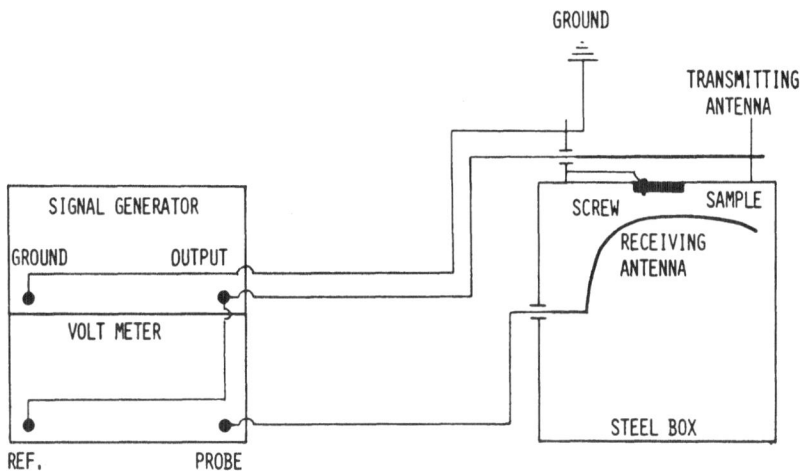

Figure 2.   Schematic of apparatus to measure EMI shielding.

EMI Shielding Materials

The device used for this measurement is shown schematically in
Figure 2.   EMI shielding is determined by the attentuation of a high
frequency signal transmitted through the test sample.   Attenuation is
expressed in decibels and is calculated from the following equation:

$$\alpha(db) = 20 \log_{10} \frac{E_t}{E_r} \quad , \qquad (2)$$

where E_t is the voltage of the transmitted signal and E_r is the voltage of the received signal.   A Wavetek Model 160 signal generator
was used to provide a signal between 1 MHz and 10 MHz.   The receiving
antenna was monitored by a Hewlett-Packard 8405 A vector voltmeter.

Thermal Measurements

Several types of thermal measurements were made.   The most basic
measurement involved determining the thermal conductivity of the composites on a hot plate device.[3]   The heat flux through the sample
was recorded by a Spectran Instrument heat flow sensor.   The surface
of the heated plate was maintained at 90 $\pm$ 1 C for all runs.

In another series of experiments the rate of temperature rise
induced by radio frequency heating was determined.   These explora-
tory studies were made primarily to assess the feasibility of using rf
heating with composites containing metallic fillers.   Since arcing

caused automatic equipment cut-off, low levels of power were used in
the study. Accordingly, nominal power levels and equipment parameters
were selected with the metal containing composites which avoided
equipment cut-off. These conditions were used for evaluation of com-
parative heating rates. Temperature rise was followed by a fine
thermocouple embedded in a 9-mm-thick compression-molded sample.

The effect of fiberglass and aluminum fibers on the curing rate
of polyester composites was also investigated. 5-cm-diameter by 15-
cm-long cylindrical castings were prepared and cured with varying
amounts of catalyst at several bath temperatures. Cycle times and
curing conditions were monitored by a thermocouple embedded in the
middle of the casting.

Tensile Property Measurements

The tensile strength and elongation to break of the samples was
measured on an Instron Tensile tester at a strain rate of 2.54/cm min.

RESULTS

Electrical Resistivity

Vulcan XC-72 carbon black was added to polycarbonate at concentrations of 11, 21, and 32 volume percent. Table 3 shows that these
samples are all quite conductive with the sample resistivity decreasing slightly from 11 ohm-cm to 2 ohm-cm with increasing carbon content. The nickel coated glass fibers were added to polycarbonate at
a total fiber loading of 43 volume percent. The resistivity of this
sample was also low at approximately 1 ohm-cm. Fiber breakage during
molding was substantial, however. The initial aspect ratio of the
fibers was approximately 450:1. After injection molding the fiber
aspect ratio was reduced to 35:1. The most detailed investigation
of composite resistivities was conducted with aluminum fibers and
flakes. Figure 3 shows the effect of aspect ratio and volume loading
on the resistivity of the composites. As the aspect ratio is increased from 12.5:1 to 24:1 the critical volume fraction decreases
from approximately 0.15 to 0.06. The lower limit of resistivity that
the composites reach is independent of the filler shape. The type
of polymer used also seems to have no significant effect on composite
resistivity for these fillers.

EMI Shielding Properties

Figure 4 shows the shielding behavior of the three carbon black
composites between 1 MHz and 10 MHz. For comparison purposes the
measured shielding response of a 3-mm-thick copper plate is also
presented. This represents the maximum response of the test facility.
The composite data which are above that of the copper plate merely
reflect experimental error. Figure 5 shows the relationship between

Table 3. Electrical Resistivity of Polycarbonate Composites

Filler	Volume loading percent	Resistivity ohm-cm
Vulcan XC-72 carbon black	11	11.0
Vulcan XC-72 carbon black	21	3.5
Vulcan XC-72 carbon black	32	2.0
Nickel coated glass fibers	43	1.0

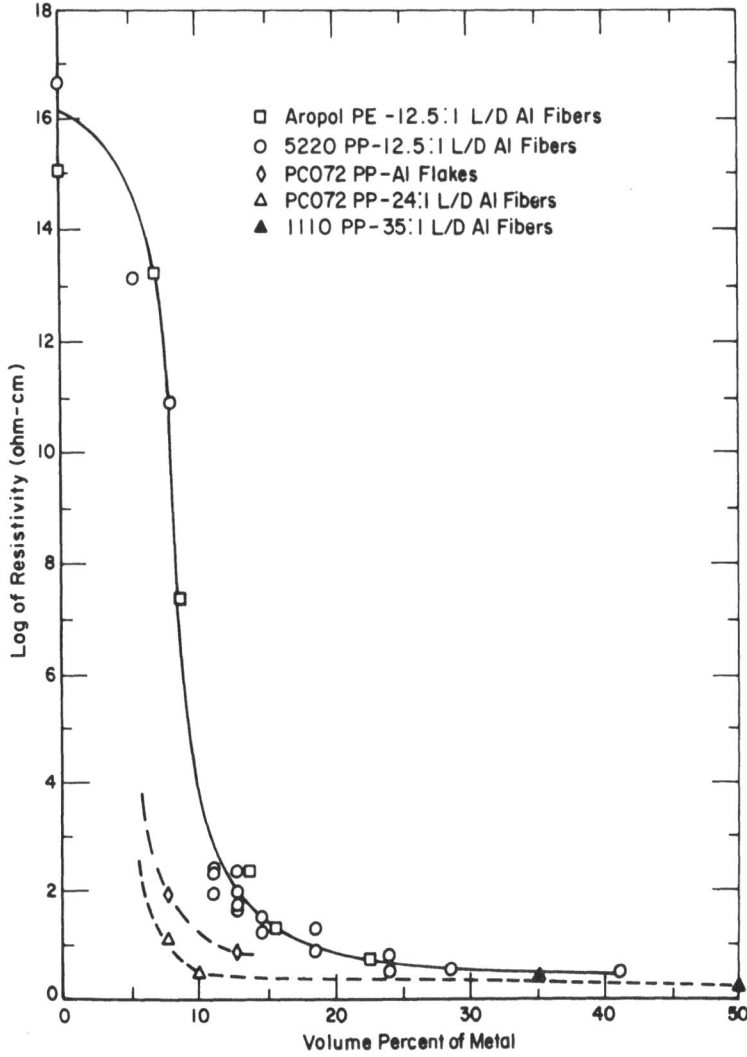

□ Aropol PE -12.5:1 L/D Al Fibers
O 5220 PP-12.5:1 L/D Al Fibers
◊ PCO72 PP-Al Flakes
△ PCO72 PP-24:1 L/D Al Fibers
▲ 1110 PP-35:1 L/D Al Fibers

Figure 3. Composite resistivity as a function of filler volume
loading and filler type.

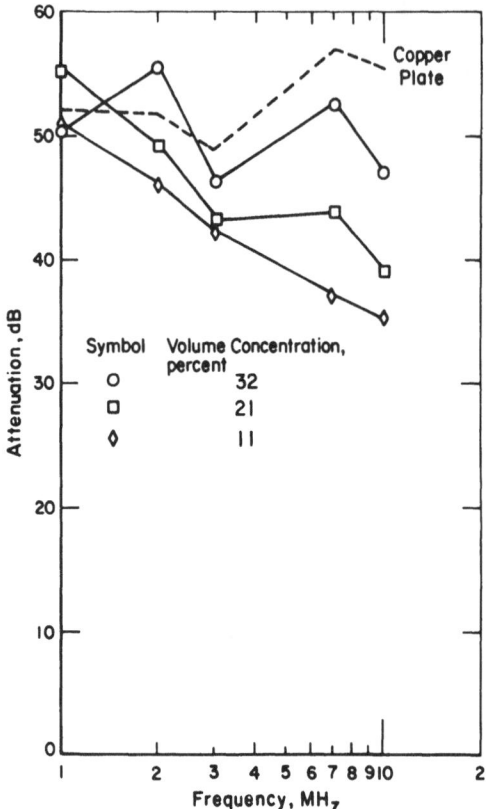

Figure 4.   Effect of concentration of Vulcan XC-72 carbon black
            on the shielding effectiveness of conductive composites.

shielding effectiveness and percent radiation transmitted. A 3-mm
copper plate should provide at least 90 dB of attentuation between
1 MHz and 10 MHz.[4] The lower recorded values of 50 dB (Figure 4)
are due to leakage around the sample edges. Since only 0.001 percent
of the transmitted radiation leaks around the copper sample, it is
apparent that where the measured response of the composites is better
than that of the metal plate, it is due to the fact that the more
compliant plastic samples may seal the aperture more effectively
than the copper plate.

     Figure 4 shows that the carbon black composites shield as well
as the copper plate at the lowest frequency studied. As the frequency
increases, however, the carbon black composites exhibit more signal
transmission than the copper plate. The loss in shielding effectiveness is related to the concentration of black, and subsequently the
resistivity of the composite. The most highly filled composite has
the lowest resistivity and the best shielding effectiveness of the
carbon black composites.

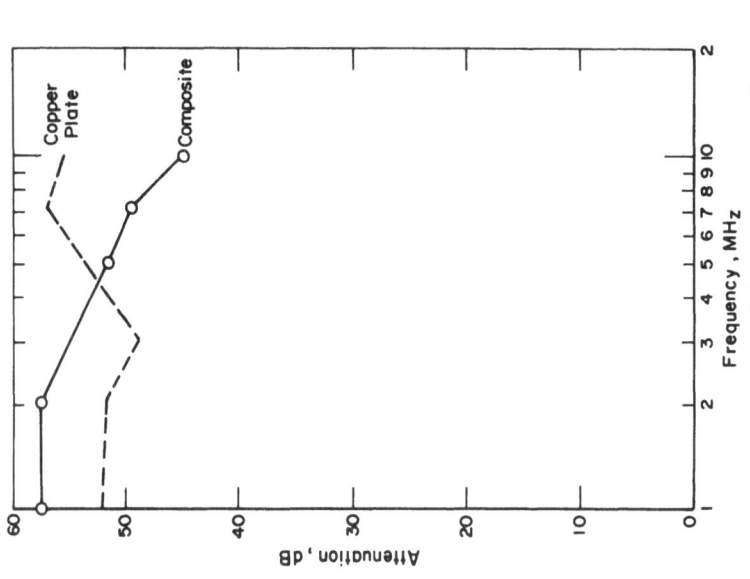

Figure 6.   Shielding effectiveness of a nickel coated glass fiber-polycarbonate composite. Fiber concentration is 43 volume percent.

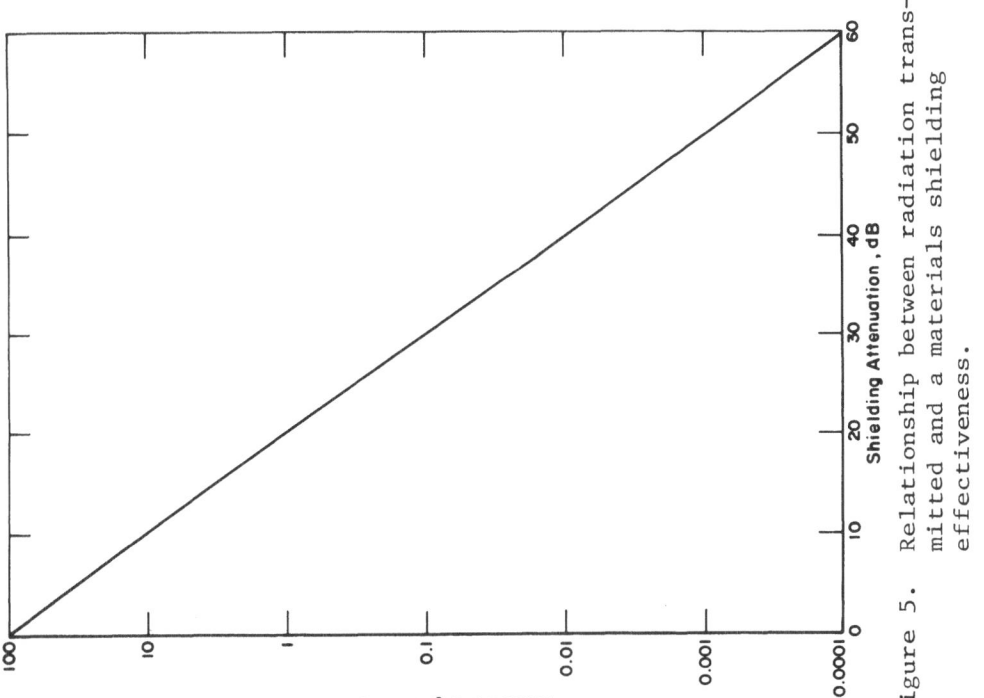

Figure 5.   Relationship between radiation transmitted and a materials shielding effectiveness.

Figure 6 shows the shielding effectiveness of the nickel coated glass fiber-polycarbonate composite. At low frequencies it shows a shielding response greater than that of the copper plate. This anomaly is most likely caused by the plastic composite's greater aperture sealing ability than the copper plate. At higher frequencies, however, more radiation is capable of getting by the composite than the copper panel. The response of the nickel coated glass fiber-polycarbonate composite is similar to the most highly loaded carbon black composite. This is consistent, since their bulk resistivities are similar.

Thermal Properties

Figure 7 compares the thermal conductivities of composites based on three types of fillers; glass fibers, copper particles, and aluminum fibers. The glass fibers used in these experiments were the milled glass fibers having an initial average aspect ratio of 240:1. After processing the aspect ratio was reduced to only 8:1, similar to the 12.5:1 of the aluminum fibers. The glass fibers increase the thermal conductivity of the composite to only two times that of the unfilled polymer at a 40 volume percent loading. The use of the more conductive copper particles increases the thermal conductivity of composites by a factor of 8 over that of the unfilled polymer. The use of aluminum fibers increases the thermal conductivity of the composites by the same factor of 8 at a volume loading of only 27 percent. Since the thermal conductivity of electrolytic copper is 4.2 watts/cm-C at 23 C, while the thermal conductivity of aluminum is only 2.2 watts/cm-C,[5] the fibrous nature of the aluminum particles has a strong influence on the thermal conductance of the composite.

The strong influence of the aluminum fibers on the thermal properties of a polymeric composite can also be seen from the data in Table 4. The presence of 3 volume percent aluminum fibers significantly reduces the induction time, overall cycle time, and peak exotherm temperature of the polyester curing reaction. The aluminum fibers used in these experiments had an aspect ratio of 35:1. Interestingly, the actual reaction period is longer when using the aluminum fibers. This is caused by the lower peak temperature which slows down the reaction. Coincident with the lower peak temperature is the fact that the temperature throughout the part is more uniform. This results in a more uniformly cured strain free part. Another consequence of having a lower peak temperature is that a higher bath, or oven, temperature can be used without causing thermal degradation to the polymer. This will lower the curing cycle time even further.

The results of the rf heating study are shown in Figure 8. Unfilled polyethylene heats slowly under the application of rf radiation. The addition of chopped glass fibers, having an aspect ratio of approximately 500:1, has little effect on the rf heating charac-

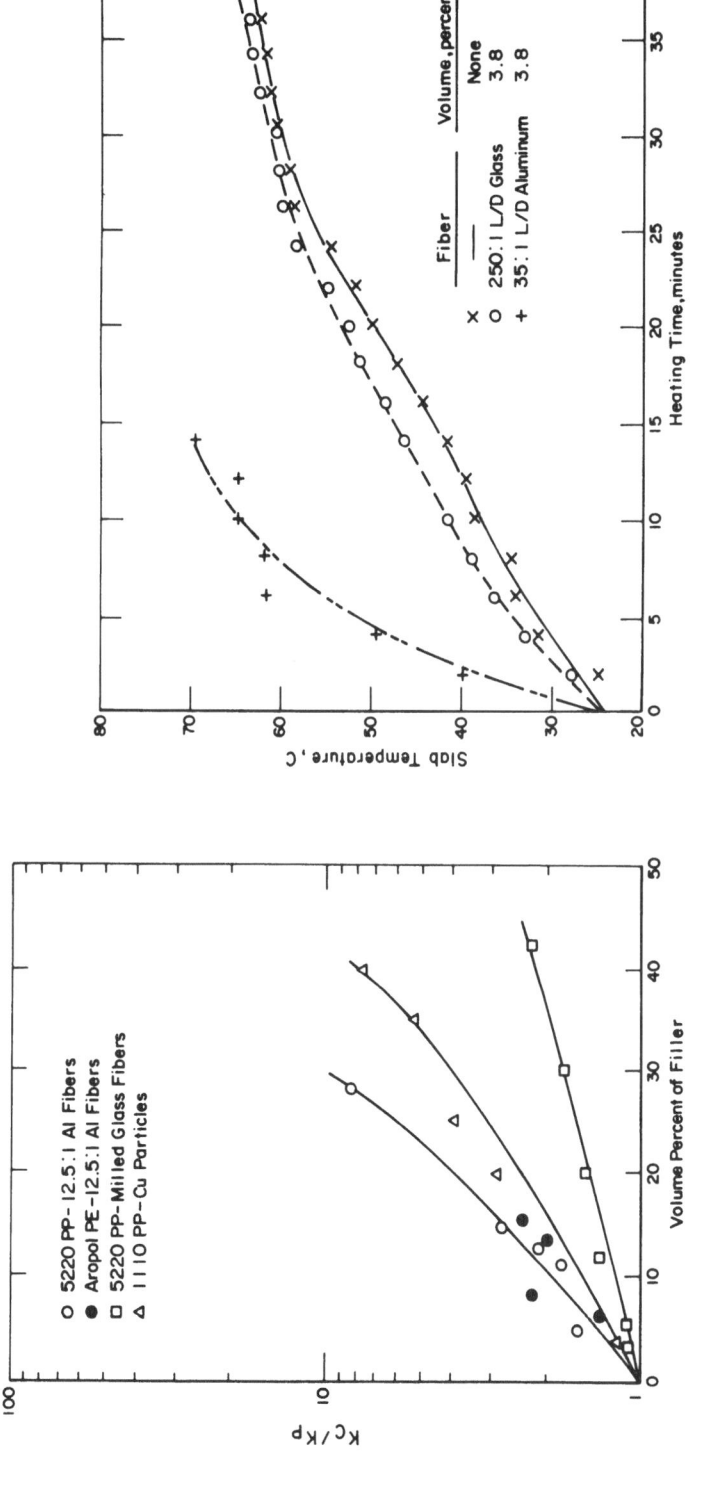

Figure 8.   Effect of fillers on rf heating rate
of polyethylene composites.

Figure 7.   Relative increase in thermal conductivity of composites as a function of
filler concentration.

Table 4. Effect of Aluminum on the Curing Rate of Polyester Composites

Catalyst concentration wt %	Glass fiber concentration vol %	Aluminum fiber concentration vol %	Bath temperature C	Induction time min	Exotherm duration min	Cool down time min	Time cycle time min	Peak exotherm temperature C
1.33	16.3	—	70	62	13	55	130	176
1.33	13.4	3.00	70	55	14	36	105	154
1.00	16.3	—	65	125	11	78	214	183
1.00	13.4	3.00	65	96	31	70	197	153
0.75	16.3	—	65	159	23	118	300	168
0.75	13.4	3.00	65	118	29	73	220	154

teristics of the material. The presence of 3.8 volume percent of
35:1 L/D aluminum fibers results in a rapid rise in the composite's
temperature.

Tensile Properties

The tensile strengths of the composites investigated are shown in
Figure 9. Data presented on this figure are averages of at least
three samples. The data show that the Shell 5220 polypropylene-
aluminum fiber composites have lower tensile strengths than the Shell
5220 polypropylene-glass fiber composites and the Hercules PCO72-
aluminum fiber composites.

The principal difference between the 5220 polypropylene and the
PCO72 polypropylene is that the latter material has been modified to
adhere better to filler particles. The data in Figure 9 reflect the
fact that general purpose polypropylene has very poor wetting properties. The much slower loss of tensile strength with increasing fiber
loading observed for the PCO72-aluminum fiber composites reflects the
improvement in adhesion between the polymer and aluminum Data for the
5220 polypropylene and the milled glass fibers behave similarly to the
PCO72 polypropylene-aluminum fiber composites. Because the aspect
ratio for all the fillers investigated is below the critical value
(fiber pullout was observed in all composites), the response of the
composites represents the degree of interfacial adhesion. The milled
glass fibers have a silane treatment to improve polymer-filler adhesion. The tensile strength data also indicate that the adhesion of the
PCO72 polypropylene to aluminum is approximately equal to that between
the silane coated glass fibers and general purpose polypropylene.

It will be noted that milled glass is not particularly effective
as a reinforcement because of the short fiber lengths. Data presented
by Conroy and Skinner for milled glass in polybutylene terephthalate
are in agreement with data presented here.[6]

There was no observed effect of particle size or geometry on the
tensile strength of the PCO72 aluminum fiber composites. Some of the
data in Figure 9 represent 50/50 mixtures of flakes and fibers. The
size range of the filler particles was not very broad, however.

The upper and lower limits to the tensile strength behavior of
filled systems has also been observed previously[7] and can, in fact,
be predicted for certain cases. Although the curves shown here are
not theoretically derived, lower curve prediction is based on the
situation whereby the polymer does not wet the filler. Conversely,
the upper curve can be predicted by assuming perfect wetting. To
date, theoretical curves have only been established for spherical
fillers. The general theoretical principles relating wetting between
a polymer and filler particles on the tensile strength of a composite
are consistent with the behavior observed in this program.

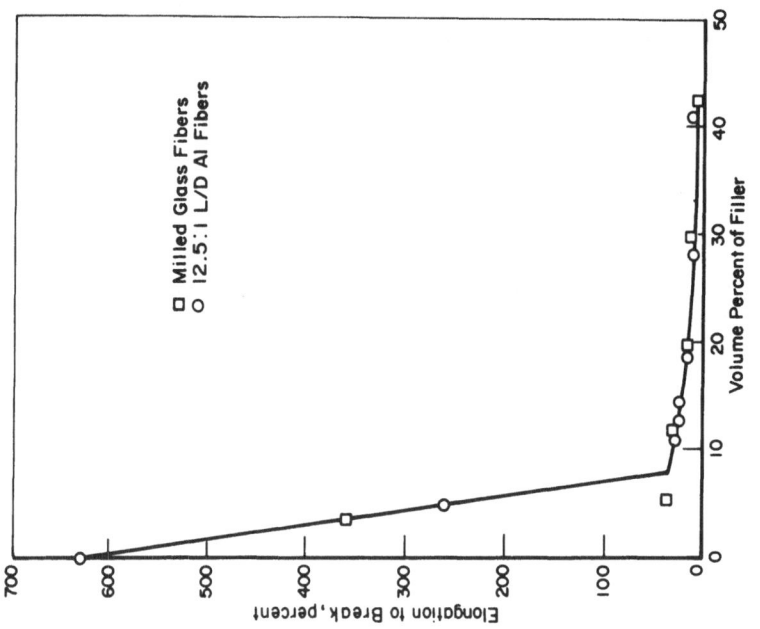

Figure 10. Effect of fillers on the fracture behavior of shell 5220 polypropylene composites.

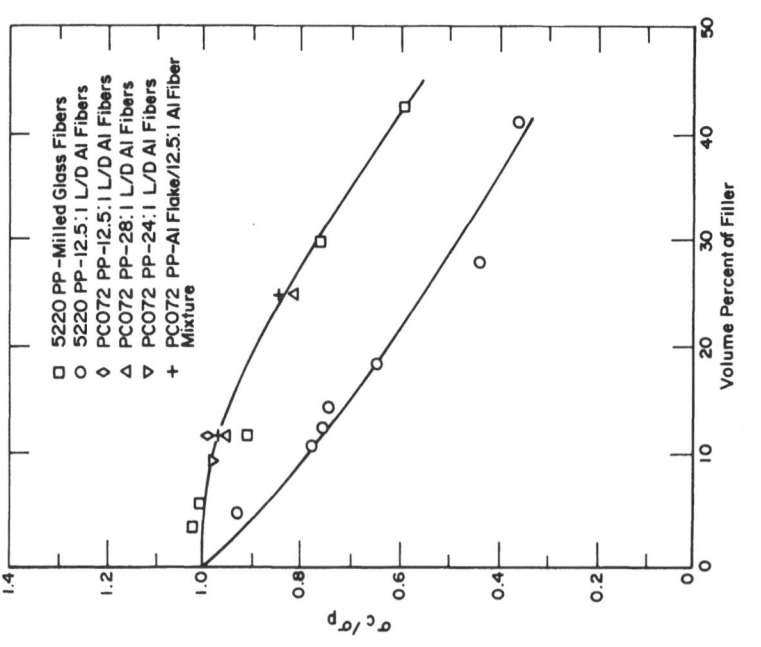

Figure 9. Relative change in tensile strength of composites with volume fraction of filler.

Figure 10 shows that the elongation to break of the aluminum fiber filled composites follows the same path as the milled glass fiber filled composites even though the wetting behavior between the two fibers and the Shell 5220 polypropylene is very different. The elongation to break drops rapidly between 0 and 8 volume percent filler loading. Above 8 volume percent the elongation to break declines gradually with increasing filler content. From this curve and Figure 9 it can be concluded that short metal fibers have the same effect on a composite's mechanical behavior as short glass fibers, providing adequate polymer-to-fiber wetting can be achieved.

SUMMARY

Because polymers are both thermal and electrical insulators it is necessary to add a conductive filler to produce conductive polymeric composites. The primary factors which influence the electrical, thermal, and mechanical properties of composites containing conductive particles are the type, shape, and concentration of filler particles. Each of the aforementioned properties is affected in a different manner by the filler particles. Composites become electrically conductive at a critical volume concentration of conductive fillers. The critical concentration can be reduced to low levels by using conductive particles which are fibrous in shape. The reduction in critical volume loading is proportional to the magnitude of the fibers' aspect ratio (length/diameter).

The thermal conductivity of polymeric composites increases monotonically with increasing filler concentration. Fibrous fillers have a dramatic effect on the magnitude of the thermal conductivity increase. This has significance in promoting more rapid and uniform curing of thermosetting resins. The use of metallic fillers also permits rf, and presumably microwave, heating to be used. These systems are useful in that the bulk item can be heated uniformly and rapidly. Because a critical volume loading is needed to induce electrical conductivity, these thermal property advantages can often be achieved without producing an electrically conductive composite.

Since the fibrous fillers that were studied were all below the critical aspect ratio for load sharing between the polymer and filler, the mechanical properties of composites using these fillers depend strongly on the degree of wetting between the polymer and filler particles. Elongation to break is not influenced by polymer-filler wetting as much as tensile strength. Under conditions of good wetting aluminum fibers can be expected to have the same effect on the composite's tensile strength as glass fibers.

The EMI shielding ability of conductive composites was shown to depend on the volume resistivity of several composites over a narrow range of transmitted frequencies. The more highly conductive a composite is the greater its ability to provide EMI shielding. It

appears that a volume resistivity on the order of 1 ohm-cm will provide 30 dB of shielding attenuation in the MHz frequency band.

REFERENCES

1. R. E. Maringer, C. E. Mobley, J. Vac. Sci. Tech., 11, p. 1067
     (1974).
2. G. D. R. Jarrett, Industrial Finishing (London), 18(218), p. 41
     (1966).
3. D. Hands, Rubber Chem. Tech., 50, p. 480 (1977).
4. R. B. Cowdell, IEEE Trans. Elec. Comp., EMC-10, p. 158 (1968).
5. R. H. Perry, C. H. Chilton, S. D. Kirkpatrick, "Chemical Engi-
     neers' Handbook", Fourth Edition, p. 12-34, McGraw-Hill,
     New York (1963).
6. A. P. Conroy, D. L. Skinner, Plast. Engr., 33, p. 28 (Aug. 1977).
7. G. Landon, G. Lewis, G. F. Boden, J. Mater. Sci., 12, p. 1605
     (1977).

METALLOPLASTICS – HIGH CONDUCTIVITY MATERIALS

D.E. Davenport

MBAssociates
P.O. Box 196
San Ramon, California 94583

INTRODUCTION

Only a few short years ago, plastics were best known for their insulating properties and great strides were made in electronics, in electrical equipment and electrical component design as the dielectric strength of plastics was improved. The whole field of thermal insulation was advanced as ways were found to generate better plastic fibers and foams which could give an order of magnitude better than ever before. And yet, today many of us are spending our time looking for ways to make plastics into better conductors – better conductors of electrons and better conductors of heat. Our reasons are as varied as the many fields from which we come, but they are all related to the fact that plastics have many other very valuable properties aside from their insulating abilities and to make full use of these in many applications, the plastics should be electrically and/or thermally conducting.

The word "metalloplastics" was created to describe an area of conductive plastics in which the conductivity was great enough to make the material resemble metal both electrically and thermally as opposed to the so called semiconductor levels achieved by adding carbon to a plastic. I'd like to describe for you the objectives of achieving such conductivities, some of the theory and experiments which we have gathered and, finally, to suggest where all of this is leading and how it may suggest new approaches – perhaps from the chemical point of view which might greatly improve on the start we have made.

THE RATIONALE FOR METALLOPLASTICS

The need for a highly conductive plastic was first seen as a way of advancing the use of plastic housings in the rapidly developing electronics industry. As computers and sophisticated electronics devices began to move out of their shielded rooms and into the cost competitive office, store and home, it became highly desirable to take advantage of the light weight, low cost and aesthetics that could be gained by subsisting plastics for metal housings for the instruments. But to provide the electro-static protection and the EMI shielding which were so vital to the many solid state components that operate at 5 volt potentials, the plastics needed a conductivity that would not only drain the static charge from the plastically dressed operator but would also provide a barrier to the radiation that was generated when the static charge was dissipated as an arc to the nearest ground. There are many stories about the troubles of the first point-of-sale instruments and electronic wrist watches which first appeared with plastic cases. It quickly became clear that plastics with a resistivity of less than 1 ohm-cm would be needed to serve these areas.

More recently, the importance of thermal conductivity in certain plastic applications has become better appreciated. As plastic housings became popular on portable tools, motors and other heat generating sources, the question of how to get the heat out of the case became an important problem. The question has become even more critical as the automobile business tries hard to use more and more plastic parts to reduce the weight of the automobile so that it will be more energy efficient. But the automobile industry is keyed to making parts in a few seconds each or a fraction of a minute at most since they must make millions a year. With conventional plastics they discovered that the problem of getting heat into the part to form it and then getting the heat out again from a poor thermal conductor meant that cycle times were in the area of minutes not seconds. This implied an increase of one to two orders of magnitude greater number of tools if they were to make the same number of parts. So, for the auto industry, an order of magnitude greater thermal conductivity could be an important advance.

For these reasons we have spent the last four years trying to understand how one can achieve large changes in the conductivities of plastics - electrical resistivities of less than 1 ohm-cm and thermal conductivities of at least a factor of ten higher than those of normal plastics. We have made significant strides toward creating this new class of metalloplastics and we have begun to understand what we must do to adapt them to conventional molding procedures and to lower their costs so they will become a practical engineering material.

THE THEORY OF METALLOPLASTICS - IMPORTANCE OF FIBER SHAPE

The basic concept that led to the creation of Metalloplastics
is that extremely small concentrations of additives can make
plastics conductive if they are in the form of conductive fibers
with length to diameter (L/D) ratios of 100 or more.

In the past, conductivity had always been obtained by the
addition of chunky fragments of copper or silver - or if only
thermal conductivity was sought, even sand particles. The results
were always the same - it required a weight percentage of 60 to 80
percent and a volume percent loading of 40 to 50 percent for
significant effects.

The striking difference between the use of chunky fragments
and fibrous materials in their effect on conductivity can be seen
from the diagram in Figure 1. If one loads a plastic with 5
percent by volume chunky fragments (shown as spheres for convenience),
a random distribution of the spheres leads to variable gaps between
the spheres, but for 6 mil spheres there will be, on the average,
a 6 mil gap to the nearest sphere. Thus, heat or electrons
flowing through such a matrix cross alternate paths of about equal
lengths in the two media.

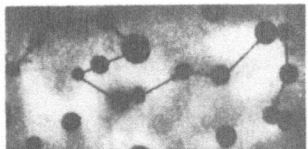

CHUNKY PARTICLES – LONG PATH THROUGH PLASTIC

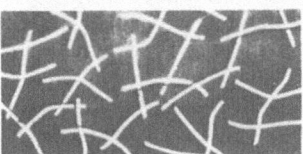

FIBERS – SHORT GAPS IN PLASTIC

Fig. 1. Comparison of a typical flow path through composites
         using the same volume percentage of material as
         spheres and fibers (L/D = 100)

If one takes the same 5 percent of material and disperses
it as 1 mil diameter fibers which are 100 mils long, the picture is
quite different. Even if the fibers were all arranged parallel
to one another and at an even spacing, they would be only 3 mils
apart. When they are allowed to take random orientations, it
becomes inevitable that they will touch one or more of their nearest neighbors, as shown in Figure 1. This provides an almost
continuous path through the composite along the highly conductive
fibers.

Thus, simple intuition tells us that fibers will be much more effective in lowering the electrical resistivity of plastic and increasing the thermal conductivity than are chunky fragments.

ELECTRICAL CONDUCTIVITY

Our intuition also tells us that effects of low concentrations of fibers on these two properties will be quite different. Whereas, thermal conductivity will be increased by such low concentrations of the fibers that one fiber does not touch its neighbors, the electrical resistivity won't be significantly modified until an almost continuous path is available through the conductive fibers.

This was clearly demonstrated with chunky fragments in 1966 by J. Garland[1]. His experimental data of electrical resistivity of Bakelite as a function of the volume percent of silver particles added is shown in Figure 2. It shows that the resistivity changes very little until the silver particle loading approaches a critical concentration - then it drops catastrophically and the composite becomes a good conductor.

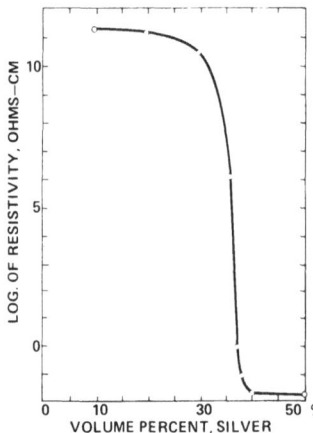

Fig. 2.   Decrease in resistivity as a function of the volume percent of chunky silver particles added to Bakelite.

Since Garland suspected the critical concentration was that associated with forming continuous chains of particles in the matrix, he devised an ingenious technique of measuring the average number of contacts each silver particle had with those around it. This data is indicated on the curve. From chain forming theory, one knows that the larger the average number of contacts, the longer the chains. Finally, when the average number of contacts approaches two, the entire particle population is interconnected in one long chain.

As Garland's data shows, the critical concentration does occur

in that narrow concentration region between the formation of significant chain lengths (m=1.26) and one continuous chain (m=2.0),
the resistivity falls relatively slowly since we are only increasing
the number of parallel paths.

When one uses fibers instead of chunky particles, the shape of
the curve remains the same but chain building starts at much lower
concentrations. In fact, one can go to statistical theory of chain
building and calculate the concentrations of fibers with a given
L/D at which a continuous chain is formed which should be a good
approximation for the onset of conductivity. The theory of course
assumes a random orientation of ideal fibers all of the same L/D
which will lead to an optimistic estimate, i.e., the lowest concentration at which one would expect to experimentally observe significant conductivity.

If one uses Milewski's[2] experimental work on fiber packing,
one can obtain a good estimate of an upper limit for the concentration at which conductivity should be observed as a function of L/D.
Milewski has shown that if one puts rigid fibers (he used wooden
dowels in his early experiments) into a volume, their packing density goes down as their L/D increases as shown in Figure 3. This
natural packing fraction results from the larger and larger voids
that are spanned by the fiber before it is supported by two or more
fibers. Furthermore, Milewski was able to show that fiberglass in
an empty container followed the same packing fraction versus L/D
curves, i.e., it behaved as rigid fibers. We have extended his
work to fibers with an L/D up to 375 using our metal coated glass
fibers which we call Metafil .

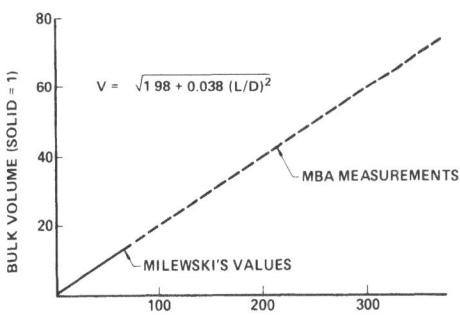

Fig. 3.   Effect of L/D on Packing Fraction.

These packing fractions are then the volume fractions required
to have two or more contacts per fiber in a matrix (average number
of contacts/fiber greater than two) and should represent the upper
limit of the volume fraction required to give conductivity. Since
this is an experimental number, it contains some non-random fibers

and inadvertent packing depending on the care with which the experiment is carried out and tends to overestimate the upper limit.

In Figure 4, we have then plotted these two estimates of the
limiting concentration along with the experimentally observed data
from various plastic processing techniques. Because the experimentally observed conductivities are with processed fibers which have
been broken down to some extent, the data are represented as bands
whose width is a best guess as to the average L/D in the matrix.

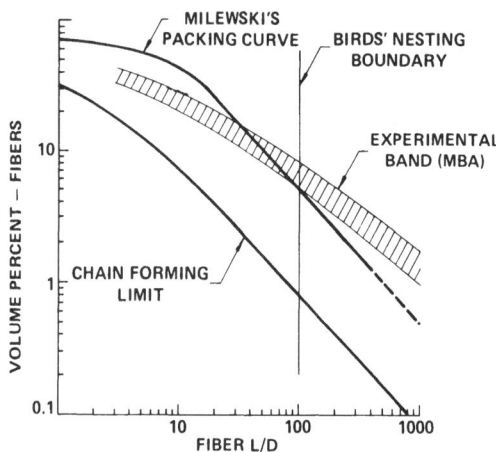

Fig. 4. Critical concentrations for conductivity versus L/D

It is interesting to note that at low concentrations, the
experimentally observed values fall between the two estimates but
that at large L/D, the experimental values fall above the upper
limit estimate. The larger deviation at the long fiber end of the
curve results from the two factors which can be readily identified.
The processing techniques to obtain data in this regime tend to give
non-random orientation because the resulting parts are thinner than
the fiber lengths. Secondly, there is a magic number for fiber tangling and matting which falls between an L/D of 50 and 100. When the
L/D values are below this critical value, the fibers pour and flow
readily without tangling. When they are above this value, the fibers
tend to tangle and form "bird's nests" which are difficult to dis-
tribute. There are ways to avoid this tangling with long fibers,
but they were not used in generating this data.

Thus, the theory and experiment confirm that to get good conductivity at very low concentrations of fibers, one should use the
maximum L/D feasible for the manufacturing process.

A puzzling aspect regarding making plastics electrically conductive with fibers is now the fibers make metal to metal contact

in the plastic matrix. One would initially suppose that there
would always be a plastic film between the fibers serving as an
insulating layer. As a matter of fact, this is probably true when
the plastic wets the fiber well and there are only low forces
between fibers. Even above the critical concentration, it has been
observed that the resistance drops markedly if the measurement voltage is greater than ten volts, implying a breakdown voltage is
being observed. To lower this interface effect, some users have
added small amounts of carbon powder. MBAssociates has developed
a process (Cross-Link™), which avoids the use of any additives
and still produces a very marked reduction in resistivity as shown
in Figure 5 over that observed with simple mixing. This figure
shows the effect of increased concentrations in lowering the resistivity for two fiber lengths and the low concentrations required
if the Cross-Link process is used. Note that resistivities as low
as 0.01 ohm-cm can be obtained at very modest fiber concentrations
which will be important for EMI shielding applications.

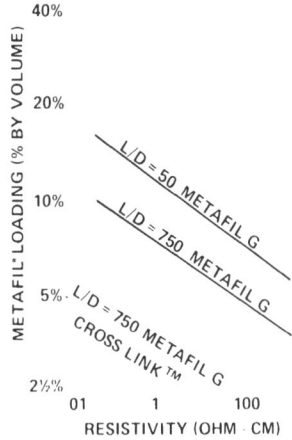

Fig. 5.   Resistance of Metafil composites – Normal and
          Cross-Link formulations.

THERMAL CONDUCTIVITY

     The thermal conductivity of a composite of high conductivity
fibers in plastics can be accurately estimated by the equation of
Nielson[3] as verified by D. Briggs of Battelle[4] with experimental
measurements on a wide range of fiber materials in plastics.

     If one inserts the appropriate parameters for Metafil fibers
into these equations, the curves shown in Figure 6 are obtained.
Unlike the electrical case, there is no critical concentration at
which dramatic changes in properties occur; rather, the curves are
all gradually changing functions of concentration. The fiber length

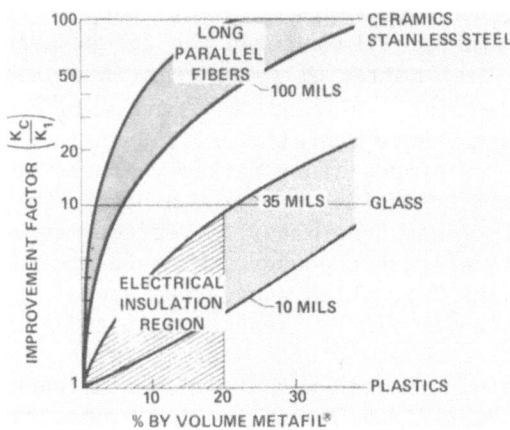

Fig. 6.   Thermal conductivity improvement when various
          Metafil lengths are added to plastic.

does play a significant role up to lengths of 0.1 inch (L/D=100);
beyond that, the change is less dramatic.

The curves show that very great increases in thermal conductivity can be achieved with low concentrations (less than 10% by
volume), particularly for the longer fibers. The upper curve for
long parallel fibers indicates that a two order of magnitude improvement in thermal conductivity is possible in the direction
parallel to the direction of the fibers, which can be very important
for composites which use unidirectional layups or fabrics.

It is to be noted that the thermal conductivity graph shows a
unique region in the low concentrations and short fiber lengths
marked "Electrical Insulation Region". This region, in which one
has no significant electrical conductivity and yet significant improvements in thermal conductivity, is to be expected since the
electrical resistivity is not reduced by low concentrations of
fibers with short fiber lengths. The region is not known to be as
sharply bounded or exactly bounded as indicated in the chart since
adequate experimental work has not been done. However, these bound-
aries are approximately correct and serve as a guide to anyone
wishing to develop plastics which are improved in thermal conductivity but can be used for electrical insulation.

FUTURE IMPROVEMENTS

As startling as these improvements are in the effect on conductivity, it should be noted that there is still room for even
more dramatic effects. The limitations to date have been in the
length of fibers which can be introduced in standard production
processes.

However, since the effects are related, not to length but to length to diameter ratios, then one may greatly improve the effects by making the fiber diameters a factor of ten or even one hundred smaller. The fibers in these studies were of the order of 1 mil in diameter but glass fibers are now routinely produced in the 0.2 mil diameter range and other materials are available in 0.01 mil diameter.

If such fibers as these could be made conductive, one would expect to get significant reduction in the concentration required for the same fiber length. The challenge would be to do this at the same low cost as can be done with the 1 mil fibers. Such fibers would withstand the high shear processing needed more readily so could even be introduced in extrusion and injection molding. Since their lengths would be much less than the wall thickness, they would also give a more nearly random orientation in the matrix.

SUMMARY

The addition of a few percent of metal and/or metallized glass fibers (L/D of the order of 100/1) to plastics leads to a family of materials which we have called "Metalloplastics". They show electrical resistivities as low as 0.01 ohm-cm so combine excellent conductivity and EMI shielding capability with the light weight and moldability of plastics. The fibers increase the thermal conductivity of the plastics up to 100 fold so can have very significant effects on molding cycle times, uniform heating and cooling rates and heat transfer rates in the final product.

Low cost, high production techniques make several million pounds of such fibers each year for other applications providing a ready supply for the plastics market. Several forms of the fibers-chopped roving, mats and fabrics – are presently undergoing testing by industry.

REFERENCES

1. "An Estimate of Contact and Continuity of Dispersion in Opaque Samples", Trans. of Met. Soc. of AIME 235, pg 642 (1966).

2. J.V. Milewski, "Micropacking: Filling Resin More Efficiently", Plastics Compounding, Volume 1, No. 1 (1978) and "A Study of the Packing of Fibers and Spheres", Ph.D. Thesis, Rutgers University (1973).

3. Industrial Eng. Chem. Fund., Vol. 13, No. 1 (1975).

4. Polymer Eng. and Science, Vol. 15, No. 12 (1977)

PLASTICS AS CURRENT AND HEAT CONDUCTORS

Robert M. Simon

Transmet Corporation
1375 Perry Street
Columbus, Ohio 43201

Historically, engineers have used metals to conduct heat and electricity, while ceramics, glass and plastics have been used as insulators. Recent developments in conductive plastics based on a number of special modifiers have allowed a unique combination of the design and processing economics of plastics coupled with the desired conductivity of metal. Applications for these unusual resin composites include electromagnetic interference (EMI) shielding, static charge dissipation, heat dissipation, and resistive heating.

As Insulators

For most applications, thermal conductivity of plastics is either disregarded or kept as low as possible. Applications in which the thermal insulating properties are plastics are used to advantage include power tools, athletic equipment, medical tools, and food service ware. This feeling of warmth to the hand due to the slow rate at which body heat is conducted away is physiologically pleasing and is frequently the major reason for the use of a plastic, i.e., tool handles, and the like.

The disadvantages of low thermal conductivity are found both in product use and manufacture. Motors which are enclosed in plastic housings frequently generate sufficient heat that the housings themselves will warp and throw bearings out of alignment, or in cases where the enclosures are non-supporting, excess heat can cause motor failure. These difficulties are present to an even greater extent in the more sophisticated electronics applications involving high power semiconductors whose resultant heat must be dumped outside the system to avoid thermal runaway. A

second drawback to thermally insulating the materials is that they
tend to dissipate heat slowly in the molding cycle. As a result,
there is excessive local heating of the polymer in the extruding
and injection molding process, resulting in some local degradation
even while the rest of the plastic is still in the solid state or a
high viscosity fluid. Then, as the melted plastic is molded in a
tool, heat must be drawn out of the system into the cold mold to
solidify the part, or in the case of thermosets, heat must be conducted into the part to cure it. Here again the thermal insulating
properties of the plastic extend the times resulting in low cycle
times. Naturally, the thicker the part the greater the handicap
is experienced.

The thermal conductivity of engineering plastics is in the
range of 4×10^{-4} to 8×10^{-4} calorie-cm/sec-cm^2-°C, while some
thermoset materials may run as high as 10^{-3}. By comparison, fused
alumina is 80×10^{-4}, soda glass is 17×10^{-4}, and asbestos paper
is 3.5×10^{-4}.

Plastics are also well known for the use as electrical insulators. Applications which began many years ago using thin mica
sheets have been replaced by thin films of polyester, polycarbonate,
and many other electrical grade resins.

Applications for electrical resistivity can take us back to
the same power tools discussed above. The switch from die cast
drill and saw housings started the "double insulated" concept, and
even allows the use of bosses as tie points for the joining of hot
electrical wires. Plugs, electrical connectors, and wire and cable
insulation all require a high degree of resistivity. Most unmodified resins exhibit bulk (volume) resistivity in the range of 10^{13}
to 10^{16} Ω cm, with a few as low as 10^{10} or as high as 10^{17}. By
comparison, glass is 10^7 - 10^{12}, copper is 1.7×10^{-6}, and aluminum
is 2.6×10^{-6} Ω cm.

As "Semi-insulators"

Plastics can be changed into partially conductive materials,
"semi-insulators", with the addition of certain modifiers or by
special treatments. Glass, while usually added as a reinforcing
agent to improve tensile strength and modulus properties of a composite, also increase the thermal conductivity. The normal range
of conductivity for base polymers is 4 to 8×10^{-4} cal-cm/sec-
cm^2-°C; 20 to 30% glass reinforced versions increase to 6 to
10×10^{-4}. Talc filled versions may show virtually no change in
thermal conductivity. Metal powders, iron and aluminum, can effectively increase heat transfer through a resin and such materials
are used at high loadings, often over 50% by weight. Applications
include grinding wheel compound and heat sinks.

Semi-insulators in the electrical area include those plastics filled with structured conductive carbon black or graphite powder, and newly developed polymers doped with modifying agents which can be incorporated directly into the polymer chain backbone itself. This class of products will usually exhibit bulk resistivity in the range of 10^4 to 10^6 Ω cm at reasonable loading levels. By comparison, silicon, a semiconductor, has bulk resistivity of 8×10^2 Ω cm.

In this range of conductivity, applications as static dissipators become viable. While base polymers could develop a large static charge through sliding contact between moving parts, the semi-insulating grades can dissipate the charge through a slow leaking process. Gears, spindles, guides, and parts of sensitive electrical equipment currently use these modified grades.

The step beyond the powdered fillers and modified resins is those resins whose property profiles are changed dramatically by the incorporation of high aspect ratio conductive modifiers. This family includes metalized glass fibers, graphite fibers, and quick quenched aluminum flakes and fibers.

In order to be used effectively, these modifiers must be amenable to the process intended. The initial high aspect ratio of a long brittle material is quickly reduced in the tortuous high shear environment of extrusion and injection molding, while the particles, short to begin with, give up the efficiency of greater aspect ratio. In all cases, efficacy of these modifiers is dependent on the aspect ratio of the particle in the final molded part.

Graphite fibers are becoming used with greater frequency in high performance applications, usually for their reinforcing properties. In the highly oriented systems such as filament winding, SMC, and layered composites, the fiber direction can be controlled to a great degree. Applications range from sporting goods, like tennis racquets, to automotive drive trains. Aircraft skin and structural components usage began with experimental military aircraft, but now include commercial aircraft components.

Although these parts may exhibit some electrical conductivity, a high amp surge from a lightning strike will damage a composite plane much more than one of an aluminum structure. For this reason a hybrid including metalized glass may gain wider use.

In the highly oriented SMC applications, a thin graphite veil changes thermal conductivity very little (thermal conductivity of graphite is 4 to 100×10^{-3}, depending on orientation) but electrical conductivity is significantly enhanced in the plane of the veil. This allows its use in some applications requiring intermediate electrical conductivity (graphite $\rho_B = 1.4 \times 10^{-3}$). An example is

an automotive SMC hood scoop in which a graphite veil is used for
shielding electromagnetic interference emitted by the engine.

Graphite fibers, both pitch and PAN, have been incorporated
into thermoplastic molding compounds for the physical property
enhancement discussed above, but since the compounding process
results in fiber breakup, dramatic improvements are seen only in
highly loaded formulations. Resins with 30% graphite fibers
exhibit high modulus (15 to 20 MPa) and enough electrical conductivity for some EMI shielding applications. The major limitation
is in cost effectiveness, as such compounds sell for about $35/kg
($15/lb).

Glass fibers metalized by vacuum deposition or dip coating
with aluminum or nickel have been tested for electrical and thermal
property enhancement. These products like the carbon fibers need
to be processed as little as possible to avoid particle breakup in
use. When employed in SMC and some BMC compounds, sufficient
aspect ratio remains to give much better thermal conductivity than
glass alone, coupled with electrical conductivity. An additional
advantage of the metalized glass is that it, like pure aluminum
particulates, is not a fine light fiber as the graphite, and thus
does not become airborne during compounding or under fire conditions
causing electrical shorts in connectors, electrical motors, etc.,
at a distance from the source.

Applications being pursued include aircraft composties in
conjunction with graphite fibers, mold heaters and EMI shielding
SMC and BMC products.

Conductive Flake Modified

The third type of high aspect ratio conductive modifiers is
based on a development by Battelle Memorial Institute to make
amorphous metal wire, ribbon, and particulates through the quick
quench (10^6C°/sec) process. While not all metal alloys processed
in this way are truly amorphous, even the crystaline forms are
micro- or crypto-crystaline in form, retaining some of the
advantages of fatigue and corrosion resistance, and ductility.
Although many alloys based in iron, nickel, copper, tin, gold,
silver, and platinum have been made, that which seems most adapt-
able for use as a filler/modifier for thermoplastic and thermoset
resins is aluminum.

The melt extraction process yields fibers currently available
from 65 to 85 microns in diameter, and 3 mm to 18 mm long. The
melt spinning technique as commercialized yields flakes 1 mm by 1.4
mm by 25-40 microns thick. Experience shows the aluminum fibers to
be of advantage in some applications which orient the fibers into a

plane, without demanding significant flow, including layup/sprayup, SMC, and mat molding.

However, those compounds which will be melt processed, thermoplastic or thermoset, and molded through the typical injection, transfer, and compression methods, should in general contain the flake product. The ductile flexible nature of the fiber tends to roll up in extensive mixing, while the flake orients itself parallel to the direction of flow, offering minimal resistance and maximum efficiency.

After the choice of product type, the property profile can be varied by concentration. Thermal conductivity increases dramatically with increased concentration at a rate much higher than with the addition of glass or mineral fillers. A 30% loading of melt extracted K-102 flake will increase the conductivity of a polymer threefold, while a 60% loading will increase it tenfold. One domestic supplier has developed a polyamide formulation with over 75% K-102. While not approaching aluminum ($\lambda = 5 \times 10^{-1}$), a phenolic measured to be 3×10^{-3} at a 35% loading would probably exhibit a λ of 10^{-2} at a 60% loading, and would form EMI shielded electronics enclosures with built in dissipation for high power applications.

Electrical conductivity in a K-102 modified composite depends on an internal network's being formed at some critical concentration. Below this concentration the polymer has enhanced thermal conductivity but is an array of conductive but non-contacting particles resulting in high electrical resistivity. When continuity is checked on a gross basis, in series with a light bulb, the light bulb will be on at full brightness or off. A carbon black or graphite fiber modified composite will cause the bulb to glow, indicating a continuous but poorly conductive path through the composite.

As with all high aspect ratio modifiers, compounding conditions, equipment and the processing time all have an effect on the efficiency of the flake and fibers. High shear, high viscosity, and long processing times are to be avoided, as they tend to shift the resistivity-to-concentration curve to the right.

Applications for resins containing melt extracted conductive modifiers include resistive heating, thermal dissipation, static discharge and perhaps the largest, electromagnetic interference (EMI) shielding. EMI shielding and electrical conductivity are interrelated in theory, and this can be found experimentally as well. Shielding effectiveness is calculated as the sum of the energy reflected by the shield and the energy absorbed by the shield. This sum is the energy loss, and is measured as a ratio of the incident wave divided by the wave passing through the shield.

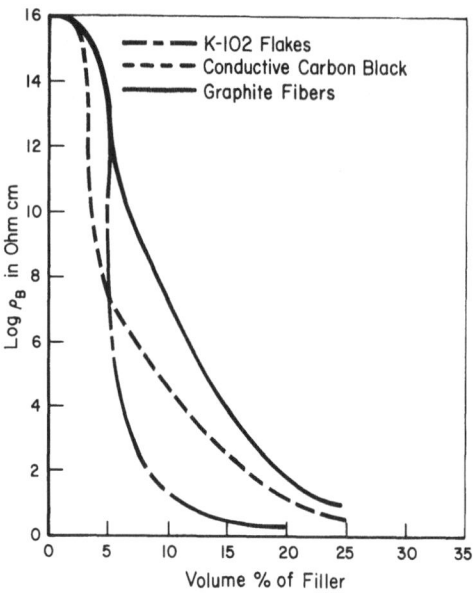

Effect of concentration on resistivity.

$$SE = R+A; \quad R = 50+10 \log (\rho_B f)^{-1}; \quad A = 1.7t \ (f/\rho_B)^{1/2}$$

$$SE = 20 \log E_1/E_2 = 20 \log H_1/H_2 = 10 \log P_1/P_2$$

Theoretical calculations based on ρ_B often give values lower than measured, due to the difficulty in measuring the conductivity of a composite. As a result, the best numbers are found by testing the resin sample itself, in a shielded box or preferably the new transmission line system developed by D. E. Stutz ("New Concepts In Shielding Using Composites", D. E. Stutz, IEEE Electro '79). The Stutz method eliminates the erroneously high shielding numbers which the boxes give due to their inherent impedence, and the erroneously low numbers given near the resonant frequency of the box. Since the various boxes alter the shielding effectiveness results, the same fixture must always be used to make comparisons. However, any system which duplicates free space measurements, like the Stutz transmission line, is immediately relatable to any other.

One advantage in the use of highly conductive plastics as inherent EMI shields is the elimination of expensive and time consuming secondary coatings which until today were necessary to protect sensitive electronics against the outside interference and to contain radiated emissions so as to pass governmental regulations.

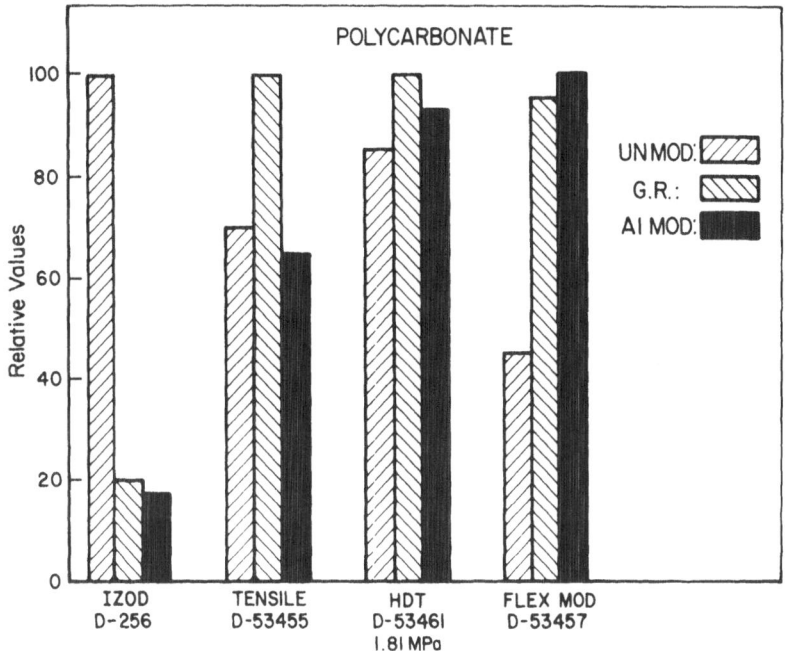

Typical Property Profile
with glass fibers and
with quick quenched flakes

The physical property profile of a K-102 modified composite
will differ from the base resin, usually ranging between glass
fiber reinforced and unmodified formulations. Experimental work
with coupling agents is already demonstrating improvement over the
unmodified flake in tensile, impact, and elongation.

During the latter part of 1980 and the first half of 1981, a
number of major resin companies are expected to release highly
conductive formulations of their base resins formulated with quick
quenched aluminum flakes. They will join the specialty compounders
who have already released product information or have compounds in
the field. These resin compositions have their own unique profiles,
in the attempt to bridge the gap between the economies of plastics
processing and the performance of metals. They do offer the design
engineer a new tool to reduce product cost and improve reliability.

SYNTHESIS AND CHARACTERIZATION OF METALLO-ORGANIC

CONDUCTING MATERIALS WITH TETRAAZANNULENES

William E. Hatfield

Department of Chemistry
University of North Carolina
Chapel Hill, North Carolina   27514

INTRODUCTION

Intensive research activities in university, industrial and national research laboratories on mixed valence organic and metallo-organic compounds have led to insights into the chemical and structural features which determine such physical properties as magnetic behavior and electrical conductivities.[1] The aim of the research is the production of new materials with properties which may be exploited in technological applications, with the expected dividends being the discovery of new phenomena, the invention of new experimental methods, and the development of new theoretical models. The recent observation[2] of superconductivity in bis(tetramethyltetraselenafulvalene)hexafluorophosphate, $(TMTSF)_2PF_6$, provides an excellent example of the successes which may be expected as this research area is expanded and developed.

An X-ray structural study[3] has shown that the nickel complex of the ligand (4,11-dihydro-dibenzo[b,i][1,4,8,11]tetraaza[14]annulene Ni(dB-TAA), is essentially planar, and that the molecules pack in slipped stacks in the solid state. Even though the slipped nature of the stacks results in a long nickel-nickel distance in adjacent molecules of 5.228 Å, the interplanar separations between crystallographically inequivalent molecules are 3.235 and 3.295 Å. These distances are slightly shorter than the 3.35 Å interplanar separation in graphite and permit extensive π-π interactions between adjacent molecules, thus making partially oxidized analogues good candidates for systematic studies of structural and chemical effects on electrical conductivities.

This paper presents a brief survey of research on mixed valence tetraazaannulene complexes which is underway in our laboratory. As noted in the reference citations much of this work is
being carried out in collaboration with research groups at the
University of North Carolina and at Texas A & M University.[4]
Systems to be discussed in this Symposium Article include a series
of macrocyclic tetraazaannulene complexes with Cu^{2+}, Ni^{2+}, Co^{2+},
Pd^{2+}, and Pt^{2+} and their partially oxidized reaction products.
These complexes will be designated as M(dB-TAA) in the following
sections. As will be shown the electrical conductivities and
activation energies of some of these compounds approach those of
the partially oxidized metallo-phthalocyanines which exhibit metal-
like temperature dependencies of σ, and of the partially oxidized
metalloporphyrins.[5]

EXPERIMENTAL SECTION

    Synthetic Procedures. - Complexes of copper(II), nickel(II),
cobalt(II), palladium(II), and platinum(II) with the tetraazaannulene were prepared[6] in good yield by an initial reaction of a metal salt (the acetate in the case of copper(II), nickel(II), and
cobalt(II); K_2PdCl_4; or K_2PtCl_4) with o-phenylenediamene in ethy-
lene glycol, followed by addition of $Na^+C_3H_3O_2^-$. The reaction
mixtures thus obtained were stirred at reflux for 5 hours, cooled
to room temperature, and the products collected by filtration.
The crude products were washed with ethanol, acetone, and diethyl
ether, and air dried. Highly purified samples were obtained by
vacuum sublimation (350°C; 0.005 mm Hg). Analytical data for carbon, hydrogen, and nitrogen on the metallocycles differed from the
calculated values by less than 0.03% in all cases.

    Oxidation with Iodine. - After refluxing for 16 hours in chlo-
robenzene in the presence of an initial ten-fold excess of iodine,
Pd(dB-TAA) yields Pd(dB-TAA)$I_{2.0}$. The mixed valence compounds
Pt(dB-TAA)$I_{1.35}$ and Pt(dB-TAA)$_{1.5}$ result from heating Pt(dB-TAA)
in 1,2,4-trichlorobenzene at 200°C in the presence of a ten-fold
excess of iodine for 16 and 24 hours, respectively. The resonance
Raman spectra of these materials using 488 nm excitation exhibited
bands at 105-115 cm^{-1} with overtones near 210 cm^{-1}, values which
are consistent with the presence of polyiodide counterions. Attempts
to obtain single crystals for structural and conductivity studies
are underway.

    Electrical Conductivities. - Electrical conductivities of
pressed pellet samples were determined by a four probe d.c. method
using the van der Pauw technique. The pellets, which were 1.3 cm
in diameter and approximately 0.2 cm thick, were compacted using
a Beckman KBr die and a ring press operated routinely at ten tons

of pressure. A Keithley Model 227 constant current source, which was operated typically at 10 µamp, and a Keithley Model 180 nanavoltmeter were used to measure the electrical conduction properties. Electrical contacts were made with silver paste. Pellets which did not provide ohmic characteristics were rejected. Activation energies were determined from temperature variation studies of the electrical conductivities in the range 77 to 300 K using a glass Dewar equipped with a sample holder which was fitted with resistance heater and temperature sensors. Temperatures were measured with a calibrated platinum resistance therometer. Measurements were made in an isothermal mode provided by the resistance heater, radiation heat leak to the cryogen, and a Lake Shore Cryotronics Model DTC-500 temperature controller. Room temperature conductivities and activation energies, where available, are summarized in Table I where it may be seen that partial oxidation of the macrocyclic complexes leads to a great enhancement of the electrical conductivities of the materials.

Table I. Electrical Conductivities and Activation Energies for $M(dB-TAA)I_x$

Compound	σ, $ohm^{-1}cm^{-1}$ (R.T.)	ΔE, eV
Cu(dB-TAA)	$<10^{-8}$	
Ni(dB-TAA)	$<10^{-8}$	
Co(dB-TAA)	$<10^{-8}$	
Pb(dB-TAA)	$\sim10^{-9}$	
Pt(dB-TAA)	$\sim10^{-8}$	
Pd(dB-TAA)$I_{2.01}$	0.4	0.04-0.06
Pt(dB-TAA)$I_{1.35}$	0.12	0.06-0.12
Pt(dB-TAA)$I_{1.50}$	0.03	

Magnetic Susceptibility Measurements. - Magnetic susceptibility data were collected using a Princeton Applied Research Model 155 vibrating sample magnetometer (VSM). The vibrating sample magnetometer was operated from zero-field to 10 kOe. The VSM magnet (Magnion H-96), power supply (Magnion HSR-1365) and associated field control unit (Magnion FFC-4 with a Rawson-Lush model 920 MCM rotating-coil gauss meter) were calibrated against NMR resonances (^{1}H and [3]Li) over the field range 0.35 - 10 KOe and found to be linear to within better than 1% over the entire range. The field set accuracy is within ±0.3% at 300 gauss and better than 0.15% at 10,000 gauss. The magnetometer was calibrated with $HgCo(NCS)_4$.[7] Powdered samples of the calibrant and compounds used in this study were contained in precision milled Lucite sample holders. Approximately 150 mg. of each were used. Diamagnetic corrections for the constituent atoms were estimated from Pascal's constants.[8-10]

Other Physical Measurements. - Electron paramagnetic resonance spectra were obtained by use of a Varian E-3 spectrometer. Infrared spectra of samples contained in KBr pellets were recorded on a Beckman IR 4250 spectrophotometer. Electronic spectra in the UV-visible range were obtained with a Cary 17I spectrophotometer on solutions of the compounds in $CHCl_3$ or Nujol mulls mounted between quartz plates. Resonance Raman spectra were taken on microcrystalline samples with 448 nm excitation. Elemental analyses were carried out by Galbraith Laboratories, Inc. of Knoxville, Tennessee.

X-Ray Diffraction Data Collection. - Diffraction data were collected on an Enraf-Nonius CAD4 automatic diffractometer with molybdenum radiation and a graphite monochromator. Accurate cell dimensions were obtained by least squares refinement of the diffractometer angle settings for 25 reflections. Procedures for the collection of intensities, calculation of scan widths, and background measurements have been described in detail elsewhere.[11] There were 3472 unique reflections for the monoclinic crystal [β-Pd(dB-TAA)] described below, of which 2361 had $F_0^2 > 3\sigma (F_0^2)$. There were 1658 unique reflections for the orthorhombic crystal [γ-Pd(dB-TAA)] of which 1001 had $F_0^2 > 1\sigma(F_0^2)$. Three dimensional Patterson analyses were used to locate the metal atoms, and the other non-hydrogen atoms were located from different Fourier maps. After further refinement of the atomic positions most of the hydrogen atoms were located from subsequent Fourier maps, with the remaining hydrogen positions being calculated. Absorption corrections were made on the data for both crystals and refinement at convergence yielded final values of R= 0.042 and R^W= 0.048 for the monoclinic crystal and R= 0.044 and R^W= 0.033 for the orthorhombic crystal. Full matrix least squares refinement was based on F, and the function minimized was $\Sigma w(|F_0|-|F_0|)^2$, where $|F_0|$ and $|F_c|$ are the observed and calculated structure factor amplitudes and $w = [2F_0/\sigma(F)]^2$.

RESULTS

X-ray crystallographic studies have revealed that there are at least three crystalline forms of Pd(dB-TAA).[11] One of these, designated α-Pd(dB-TAA), which was obtained by vacuum sublimation of the crude reaction product crystallizes in the monoclinic system with a = 19.426(23), b = 5.292(7), c = 14.838(3), and β = 112.32(8). Based on the similarity in the structural data for α-Pd(dB-TAA) and Ni(dB-TAA) which is summarized in Table II, it is reasonable to conclude that the two compounds are isostructural, with planar α-Pd(dB-TAA) molecules stacking in a slipped fashion.

A second crystalline form of the palladium compound, β-Pd(dB-TAA), was obtained from the batch of crystals which yielded α-Pd-(dB-TAA). The structure

Table II Unit Cell Data for Ni(dB-TAA) and α-Pd(dB-TAA)

Compound	a, Å	b, Å	c, Å	β, deg
Ni(dB-TAA)[a]	19.456(4)	5.228(1)	14.868(3)	112.28(1)
α-Pd(dB-TAA)[b]	19.426(23)	5.292(7)	14.838(3)	112.32(8)

[a]Reference 4.

[b]Reference 11.

of Pd(dB-TAA) has now been solved by X-ray crystallography. A view of a single molecular unit is given in Figure 1, and some important molecular dimensions are given in Table III. The material crystallizes in the monoclinic system, space group $P2_1/c$ with four molecules in a unit cell of dimensions $\underline{a}$ = 8.974(8), $\underline{b}$ = 10.985(10), $\underline{c}$ = 14.617(10), and β = 95.14(9). Although all atoms sit on general positions, the molecule is very nearly planar, with the largest deviation from the best least squares plane for all non-hydrogen atoms being 0.047(.010)Å for an apical carbon (C11) of one of the propane-1,2-diiminato exocyclic rings. The palladium atom lies at -0.035(.001)Å with respect to this plane. In a similar calculation, the palladium atom sits at -0.016(.001)Å from the best least squares plane formed by the four nitrogen donor atoms.

As shown in Figure 2, the packing of the molecules in the solid state is distinctly different from that exhibted by the nickel analogue. Instead of the slipped molecular stacking as occurs in Ni(dB-TAA), the molecules of β-Pd(dB-TAA) are arranged in pairs. The shortest palladium-palladium distance is 4.382(1)Å, with the next shortest distance between palladium atoms in adjacent pairs being 6.643 Å. The closest contact of the palladium ion with an adjacent molecule of a given pair is to C9 with Pd-C9' being 3.432-(4) Å. Other close contacts are Pd-N2' of 4.178(4) Å and Pd-N3' of 3.645(4) Å. The Pd-C9'-Pd' angle is 88.02(11)°.

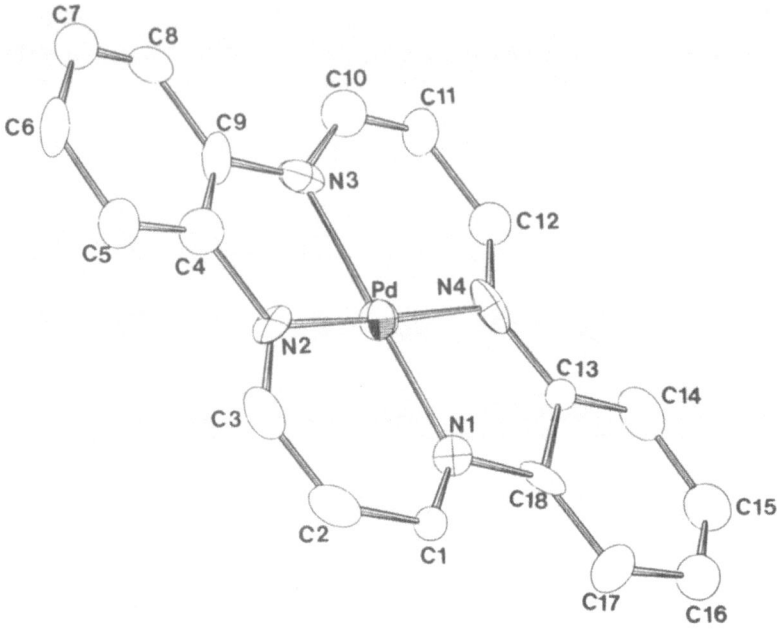

Figure 1.   A view of the structure of Pd(dB-TAA).

The average bond distance in the C4-C9 benzenoid ring is
1.390 Å, and the average angle in the ring is 120.0°. The comparable average bond distance and angle in the C13-C18 ring is
1.397 Å and 120.0°. The internuclear distances between the nitrogen donor atoms in the coordination plane are

N1-N2	2.915(5) Å
N1-N4	2.609(5) Å
N2-N3	2.605(5) Å
N3-N4	2.938(5) Å

In an attempt to find an adequate crystal for a crystallographic study of α-Pd(dB-TAA), a third crystalline form of Pd(dB-
TAA) was found. This third form, γ-Pd(dB-TAA), crystallizes in the
orthorhombic system, space group $Pn2_1a$ with the cell dimensions a =
14.793(3), b = 18.001(6), and c = 5.291(3) Å, and Z = 4. The
gross features of the structure of γ-Pd(dB-TAA) are similar to

those of β-Pd(dB-TAA), and the corresponding bond distances and angles are compared with those of β-Pd(dB-TAA) in Table III, where the atom numbering scheme given in Figure 1 is used. The estimated standard deviations for the bond distances and bond angles are somewhat larger than we usually require, but the number of data were limited and the results are adequate for the purposes of this study.

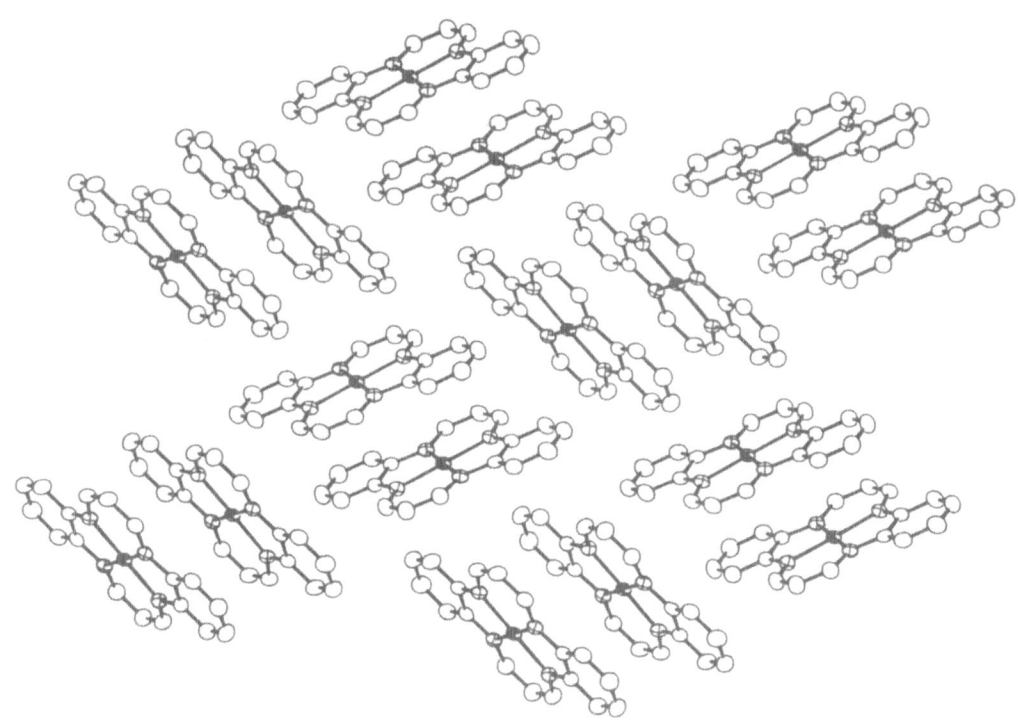

Figure 2. *A view of several β-Pd(dB-TAA) molecules projected on the bc plane which shows the formation of cofacial dimers and the packing of the dimers.*

Table III. Selected Bond Distances and Angles for β-Pd(dB-TAA)
            and γ-Pd(dB-TAA)

Bond Lengths, β-Pd(dB-TAA		Bond Lengths, γ-Pd(dB-TAA)
Pd–N1	1.959(3)	1.925(10)
Pd–N2	1.949(4)	1.935(15)
Pd–N3	1.967(3)	2.011(11)
Pd–N4	1.962(4)	1.984(15)
N1–C1	1.323(5)	1.280(16)
N1–C18	1.422(6)	1.445(14)
C1–C2	1.384(6)	1.498(19)
C2–C3	1.385(6)	1.424(18)
N2–C3	1.328(5)	1.366(21)
N2–C4	1.430(5)	1.437(29)
N3–C9	1.418(6)	1.386(15)
N3–C10	1.318(5)	1.405(19)
C10–C11	1.390(7)	1.250(20)
C11–C12	1.397(7)	1.404(16)
N4–C12	1.309(5)	1.295(22)
N4–C13	1.415(5)	1.376(29)

Bond Angles, β-Pd(dB-TAA)		Bond Angles, γ-Pd(dB-TAA)
N1–Pd–N2	96.41(15)	97.59(.69)
N2–Pd–N3	83.37(15)	81.41(.71)
N3–Pd–N4	96.80(15)	96.66(.79)
N4–Pd–N1	83.41(15)	84.31(.80)
Pd–N1–C1	122.23(32)	127.80(.99)
Pd–N1–C18	113.53(28)	113.12(.78)
Pd–N2–C3	122.99(32)	121.86(1.30)
Pd–N2–C4	113.21(28)	115.30(1.35)
Pd–N3–C9	113.46(28)	112.19(.82)
Pd–N3–C10	121.71(32)	115.84(.98)
Pd–N4–C12	122.80(32)	121.44(1.63)
Pd–N4–C13	113.36(29)	112.68(1.32)
C1–C2–C3	128.21(44)	126.63(1.01)
C10–C11–C12	128.61(43)	126.91(1.34)

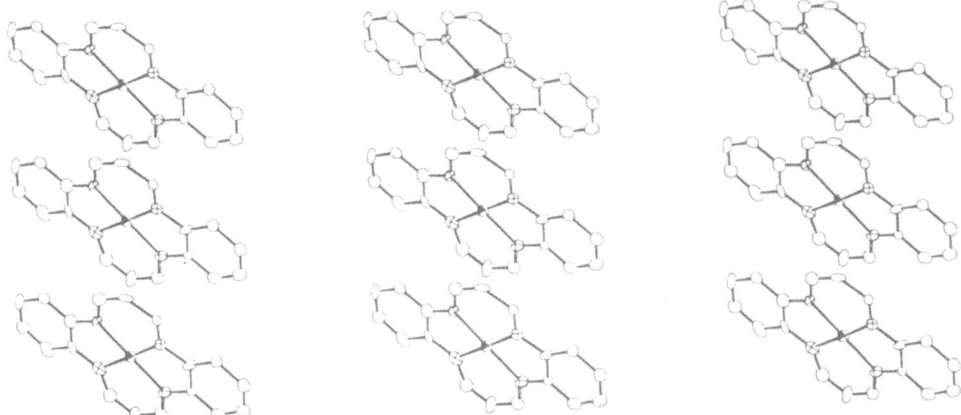

*Figure 3.   A projection of a segement of the structure of
            γ-Pd(dB-TAA) on the ac plane which shows the stacking
            arrangement of the molecular units and the relative
            orientation of the chains.*

As shown in Figure 3, the molecular units in γ-Pd(dB-TAA)
stack to form chains which are intimately packed in the solid
state.   Some close contacts between a given molecule and one
related by one translation unit along the z axis (designated by
primes) are

Pd–C$_2$′	3.797(14) Å	N1–N4′	3.899(17)
Pd–C$_3$′	3.541(11) Å	N1–C2′	3.811(12)
Pd–C11′	3.804(16) Å	N1–C12′	3.835(15)
Pd–C12′	3.537(11) Å	N1–C13′	3.756(16)
		N1–C14	3.473(15)

The C$_2$–Pd–C$_2$′ angle is 97.20(.3)°.   These distances emphazise the
slipped nature of the stacking arrangement, which may arise from
enhanced van der Waals attractive forces permitted by the stacking
pattern.   Molecular orbital calculations which are underway on one
pair of molecules with the geometry exhibited by β-Pd(dB-TAA) and
a second pair with the geometry in γ-Pd(dB-TAA) are underway and
the results may lead to an explanation for the observed packing
arrangements.

DISCUSSION

The observation of three crystalline forms of the precursor compound Pd(dB-TAA) is extremely interesting since all three crystals came from the same batch of material that had been purified by vacuum sublimation. One may conclude that the different crystal forms resulted from crystal growth along the thermal gradient of the cold finger in the sublimation apparatus.

The oxidation of Pd(dB-TAA) with iodine in refluxing chlorobenzene yielded a black solid with stoichiometry $Pd(dB-TAA)I_{2.01}$. The resonance Raman exhibited a strong absorption at 105 cm^{-1} with an overtone at 212 cm^{-1}. The 105 cm^{-1} absorption band has been identified as being characteristic of the totally symmetric I-I-I stretching motion.[12] This observation provides evidence for the formulation $Pd(dB-TAA)(I_3)_{0.67}$.

The electrical conductivity of the precursor compound Pd(dB-TAA) is less than 10^{-9} ohm^{-1} cm^{-1}. Upon iodination there is a dramatic increase in conductivity with the conductivity of the partially oxidized product being 0.4 ohm^{-1} cm^{-1}.

The EPR spectrum of $Pd(dB-TAA)I_{2.01}$ shows a broad singlet at g= 2.004, and magnetic susceptibility measurements confirms the near diamagnetism expected for the band-like electronic structure. As expected the bands in the infrared spectrum of the partially oxidized material are obscured.

The electronic spectrum of Pd(dB-TAA) in $CHCl_3$ differs from that of a Nujol mull of a sample of the material in that the lowest energy band (at 506 nm) is not present in the solution spectrum. This band is likely a result of the stacking arrangement in the solid state. There is an absorption band at 410 nm in Pd(dB-TAA) which is unaffected by iodination. This band is tentatively assigned to a metal-ligand charge transfer transition but the assignment can be made more definite once the molecular orbital calculations are completed. An absorption due to the polyiodide could not be made with certainty.

Depending on the length of the reaction time, oxidation of Pt(dB-TAA) with iodine in near boiling 1,2,3-trichlorobenzene resulted in two complexes with different iodine contents. After 18 hours a material with the composition $Pt(dB-TAA)I_{1.35}$ was obtained, while extending the reaction time to 24 hours afforded a material with a higher iodine content, that being $Pt(dB-TAA)I_{1.5}$. Since the resonance Raman of this latter material shows absorption bands at 114 cm^{-1} and 174 cm^{-1}, it is most reasonable to conclude that the polyiodide is a highly distorted I_3^- or I_5^- moiety.[12]

Iodination of Pt(dB-TAA) leads to a great increase in the electrical conductivity. Of the two samples, the one with the lower iodine content exhibited the higher conductivity at room temperature. The electrical conductivity of Pt(dB-TAA)$I_{1.35}$ is 0.12 ohm^{-1} cm^{-1} (pressed pellet) while that of Pt(dB-TAA)$I_{1.5}$ is 0.03 ohm^{-1} cm^{-1}.

The room temperature EPR spectra of the iodinated platinum samples consisted of one broad band, respectively, at g= 2.003, and magnetic susceptibility measurements on Pt(dB-TAA)$I_{1.35}$ demonstrated that the material was nearly diamagnetic, as expected.

As in the case of Pd(dB-TAA), the low energy transition at 560 nm in the Nujol mull spectrum of Pt(dB-TAA) is the characteristic feature of the stacked structure in the solid state. This transition is either shifted to 625 nm upon iodination, or a new band appears in the Nujol spectrum. A strong absorption band at about 410 nm is unaffected by iodination. The spectrum of Pt(dB-TAA)$I_{1.35}$ is similar to that of Pt(dB-TAA)$I_{1.5}$, with the exception that a strong band appears in the spectrum of Pt(dB-TAA)$I_{1.35}$ at about 720 nm. This latter band may be due to the polyiodide, but the resonance Raman spectrum of Pt(dB-TAA)$I_{1.35}$ is unavailable at this present time, and a conclusion must await the completion of the experimental work.

CONCLUDING REMARKS

Since it is well known that electrical conductivities of pressed pellets are considerably smaller than the intrinsic conductivities of the materials because of contributions to the resistivity from interparticle contact resistances and because of the inherent anisotropy of low-dimensional systems, there is considerable stimulation for continued efforts to obtain single crystal samples of M(dB-TAA)I_x. It is likely that such single crystals will exhibit metal-like conductivities i.e. dσ/dT<0, since the conductivities and activation energies of pressed pellet samples of these materials are comparable to those observed for Ni (phthalocyanine)(I$_3$)$_{0.3}$, a substance known to exhibit metal-like conductivity.[5b]

The relatively long palladium-palladium distances in the stacked and dimeric forms of Pd(dB-TAA) leads to the conclusion that metal-metal interactions do not contribute significantly to any band formation, and that intermolecular interactions occur between the π-orbitals of partially oxidized macrocyclic complexes.

ACKNOWLEDGEMENT

This research was supported in part by the Office of Naval Research. I am grateful to my colleagues for their continued interest and collaboration in this research.

REFERENCES

1. Current research activities are summarized in the following proceedings:
   a. Baristic, S.; Bjelis, A.; Cooper, J.R.; Leontic, B. "Quasi-One-Dimensional Conductors"; Notes in Physics; Springer-Verlag:Berlin, 1979; Vol. 95 96.
   b. Hatfield, W.E. "Molecular Metals"; NATO Conference Series VI; Plenum: New York, 1979; Vol. 1.
   c. Miller, J.S.; Epstein, A.J. "Synthesis and Properties of Low-Dimensional Materials"; Anal of the New York Academy of Sciences; New York Academy of Sciences, 1978; Vol. 313.

2. Jerome, D.; Mazaud, A.; Ribault, M.; Bechgaard, K. J. Phys. Letters 1980, Feb. 15.

3. Weiss, M.C.; Gordon, G.; Goedken, V.L. Inorg. Chem. 1977, $\underline{16}$, 305.

4. M. Tsutsui and R. Boxsein prepared samples of a number of these compounds and the UNC Group carried out physical measurements. This joint research will be described in detail elsewhere.

5. a. Peterson, J.L.; Schramm, C.J.; Stojakovic, D.R.; Hoffman, B.M.; Marks, T.J. J. Am. Chem. Soc. 1977, $\underline{99}$, 286.
   b. Schramm, C.J.; Stojakovic, D.R.; Hoffman, B.M. Marks, T.J. Science 1978, $\underline{200}$, 47.
   c. Wright, S.K.; Schramm, C.J.; Phillips, T.E.; Scholler, D.M.; Hoffman, B.M. Synthetic Metals 1979/80, $\underline{1}$, 43.

6. Corvan, P.J.; Lau, C.P.; Hatfield, W.E. manuscript in preparation.

7. Brown, D.B.; Crawford, V.H.; Hall, J.W.; Hatfield, W.E.; J. Phys. Chem. $\underline{1977}$, $\underline{81}$, 1303.

8. Figgis, B.N.; Lewis, J.; In "Modern Coordination Chemistry", Lewis, J. And Wilkins, R.G., Editors, Interscience Publishers, Inc.: New York, 1960; Chapter, 6; Page 403.

9. König, E., "Magnetic Properties of Transition Metal Compounds", Springer-Verlag: Berlin; 1966.

10. Weller, R.R.; Hatfield, W.E.; J. Chem. Ed. 1979, 56, 652.

11. Hatfield, W.E.; Hodgson, D.J.; Brookhart, M.; Corvan, P.J.;
    Lau, C.F.; Marsh, W.E.; Boxsein, R.; Tsutsui, M. manuscript
    submitted for publication.

12. Cowie, M.; Gleizes, A.; Grynkewich, G.W.; Kalina, D.W.;
    McClure, M.S.; Scaringe, R.P.; Teitelbaum, R.C.; Ruby, S.L.;
    Ibers, J.A.; Kannewurf, C.R.; Marks, T.J. J. Am. Chem. Soc.
    1980, 101, 2921, and references therein.

ELECTRICAL AND PHYSICAL PROPERTIES OF THIN FILMS CONTAINING

CHARGED MICROGEL POLYMERS

Donald A. Upson and Gerald A. Campbell

Research Laboratories, Eastman Kodak Company

Rochester, New York 14650

INTRODUCTION

Conductive polymeric thin films are applicable for controlling the accumulation of static electricity on insulating substrates. Historically, water-soluble polyionomers mixed with
hydrophilic binder polymers have been used for this purpose.[1] The
functions of the binder are to provide flexibility, improve
adhesion, impart a degree of permanence, and act as a moisture
sump. This latter function has been considered important, since
the conductivity of polyionomers is dependent on relative humidity.[2] Such conductive films generally possess certain physical
property deficiencies, however, including poor abrasion resistance
and tackiness at high temperatures and relative humidities. If
soluble polyionomers are coated with a hydrophobic binder, tackiness at higher temperatures and humidities is still observed,
indicative that the polyionomer still has a significant effect on
the physical properties of the coating surface.

In order to provide excellent levels of conductivity in thin
films that do not depend on hydrophilic binders for water reten-
tion, highly charged polymeric microgels have been developed.
These microgels are conductive in the presence of hydrophobic
binders and have a minimal effect on the physical properties of
the binder.

MATERIALS

Charged microgel polymers are crosslinked spherical poly-
electrolyte particles dispersible in solvents in which non-cross-
linked polymers of similar composition would be soluble. The

polymers are 20-99 mole % ionic and are in the colloidal size
range of about 0.01 μm to 1.0 μm. Both highly anionic and highly
cationic microgels have been prepared.[3,4] Structurally, the
microgels take the following form:

$$-(A)_x---(B)_y---(CH_2-CH)_z---$$

L
|
M   X

Moiety A can be any copolymerizable monomer; B is a di- or
multi-functional monomer capable of crosslinking the polymer; L is
a single bond or a linking group; M is a charged ionic group
covalently bound to L; and X is a counterion. Mole percentage x
is generally less than 20, y is generally 5-20, and z is 60-95,
although other ratios have also been found useful. These polymers
can be conveniently prepared by emulsion copolymerization of
monomers A and B with another water-insoluble monomer which can be
transformed by a post-polymerization reaction to an ion-containing
species. For cationic microgels, vinylbenzyl chloride is copolymerized as described above. No isolation or purification of this
precursor latex is necessary. The crosslinked latex is treated
with a tertiary amine or phosphine to form in situ the corresponding quaternary ammonium or phosphonium chloride sites. For
anionic microgels, a vinyl aromatic monomer is copolymerized, and
the crosslinked latex is isolated, dried, dispersed in a chlori-
nated organic solvent, sulfonated with complexed sulfur trioxide,
isolated, and redispersed in a desired solvent. Both procedures
produce highly charged microgels having size dimensions essentially identical with those of the precursor crosslinked hydrophobic latex.

The microgels can be isolated, if desired, by precipitation
in nonpolar organic solvents, yet readily redispersed in water,
methanol, or other polar solvents. This versatility permits wide
formulation latitude. Since binders are often used to improve
adhesion and to confer permanence to the film, polymers that are
dispersible in water (e.g. latices) or are soluble or dispersible
in relatively polar organic solvents can be considered.

RESULTS AND DISCUSSION

Microgels combined with latex polymer binders or with or-
ganic-solvent-soluble polymeric binders have given the most satis-

factory results. The choice of a binder hinges on compatibility with the conductive polymer and with the solvent system. When these criteria are satisfied, the binder polymer can then be chosen or designed based on the physical property requirements of the specific application. The physical properties observed are nearly identical with those of the binder polymer alone with little effect of the microgel other than conferring conductivity. For example, two-component films wherein the ratio of binder to conductive polymer is one or greater are not softened by moisture at high relative humidities and temperatures. The result is an ionically conductive film that does not stick or become tacky at any normally observed set of atmospheric conditions. In addition, the films exhibit abrasion resistance nearly equivalent to that of the binder polymer alone. As ratios of binder to microgel are lowered below one, progressive compromise of physical properties is observed. At ratios more than about eight, conductivity is compromised. Optical clarity in the two-phase coating is achieved by matching the refractive indices of the conductive polymer particles and the binder to within about 0.04 units. The thickness of the coatings is generally 0.2-1.0 μm. Below 0.2 μm, microgel coverage is insufficient for good conductivity. Little advantage in terms of either conductivity or physical properties accrues for coatings thicker than 1.0 μm.

The surface electrical resistivity values of microgel-containing films are dependent on microgel coverage and relative humidity, as shown in Figures 1 and 2. The responses shown are typical for polyelectrolytes.[2]

The hydrophobic polymers useful as binders in many cases are found toward the positive end of a triboelectric series. The practical significance of this property is that these materials, on contact and separation with dissimilar surfaces, are more likely to acquire static charge than other materials that occupy a more neutral position in the triboelectric series.[5] Cationic microgels aggravate the positive charging characteristics of hydrophobic binders. Although this effect is of minor importance in a highly conductive film, competition is established between the rate of electrification and the rate of dissipation. In very high-speed operations, static buildup can occur despite the presence of the conductive agent. Although the anionic polymers are less conductive than their cationic counterparts, the anionic microgels contribute toward negative triboelectric charging. When anionic microgels are coupled with a hydrophobic binder, the net effect is a more nearly neutral charging film that has reduced tendency to acquire static charge on contact and separation with other surfaces.

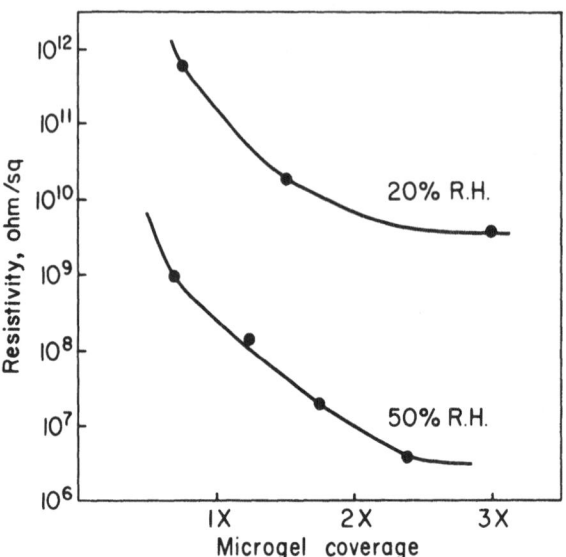

Fig. 1. Cationic microgel. Coverage vs.
         resistivity.

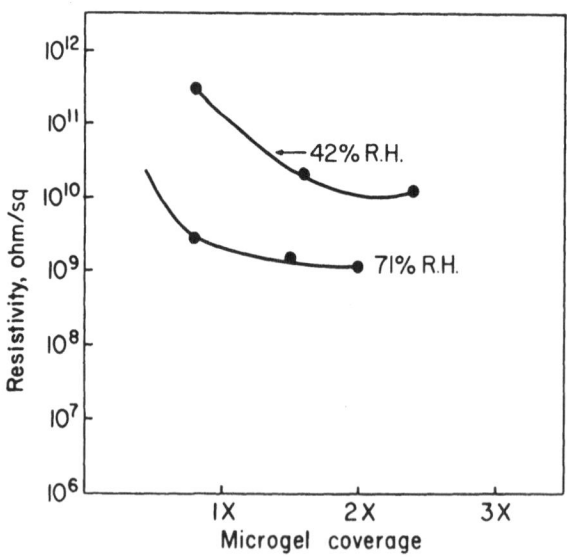

Fig. 2. Anionic microgel. Coverage vs.
         resistivity.

CONCLUSION

Both cationic and anionic charged microgel polymers have been prepared which are useful as conductive agents in thin films. They each can be combined with appropriate binder polymers. The physical properties of the conductive thin film are very similar to those of the binder polymer alone with minimal effect by the microgel on properties other than conductivity. Of the two microgels, the cationic version is more conductive, but the anionic version has less tendency to acquire triboelectric charge.

REFERENCES

1. K. Johnson, "Antistatic Compositions for Textiles and Plastics," Noyes Data, Park Ridge, N.J. (1976) p. 280-323.
2. W. C. Meyer, Tappi, 57(3):86-88 (1974).
3. R. N. Kelley and G. A. Campbell, US patent 4,070,189.
4. G. A. Campbell and R. N. Kelley, US patent 4,147,550.
5. M. W. Williams, J. Macromol. Sci.-Rev. Macromol. Chem., C14(2):251-265 (1976).

ELECTRICAL PROPERTIES OF GROUP IV B METALLOCENE POLYOXIMES

Charles E. Carraher, Jr., Raymond J. Linville, Tushar A. Manek, Howard S. Blaxall, J. Richard Taylor, Larry P. Torre

Department of Chemistry
Wright State University
Dayton, Ohio 45435

INTRODUCTION

The synthesis and characterization of organometallic polymers is an ongoing portion of our research program. Recently we reported the synthesis of polyoximes of form I when M = Ti . More recently we synthesized the analogous Group IV B polyoximes where M = Zr and Hf .

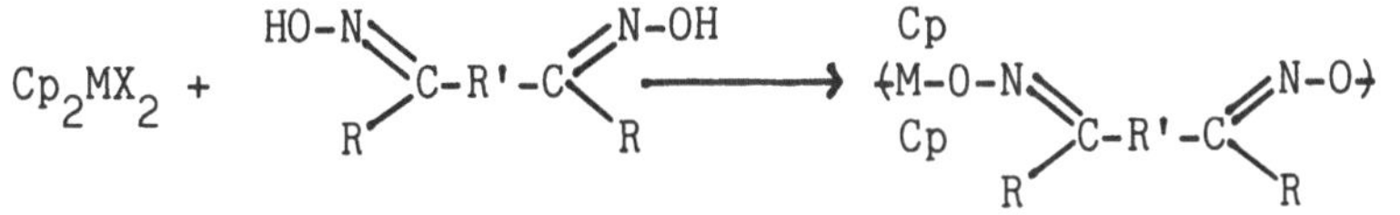

The reasons for us desiring the synthesis of such polyoximes includes evaluation of their electrical properties. Here we report preliminary studies on several of these polyoximes.

EXPERIMENTAL

The polyoximes were synthesized as described in reference 1. Bulk DC resistivities were obtained using a RCA Model WV-S11A Picoammeter and Keithley 173A Multimeter coupled through a custom-built resistivity cell to a Hewlett-Packard Model 6516A DC power supply or ELCO 1030 power supply. The Keithley 173A Multimeter was utilized to both measure amperage and the utilized voltage thus providing a "double check" system. Bulk specific resistance is calculated from the relationship $p = Ra/t$ where R is the resistance (ohms), a is the surface area of the pellet (cm^2), t is the thickness of the pellet (cm) and p is the bulk specific resistance (ohm-cm).

Bulk capacitance, K, and power dissipation factor, D, values were obtained using a component-built AC bridge consisting of a General Radio Model 1311-A audio oscillator (as an external generator) connected to a General Radio type 1608-A impedance bridge using a General Radio type 1232-A tuned amplifier and null detector. The bridge was coupled through a custom-built sample cell.

The dielectric constant is a ratio of the capacities of a parallel plate condenser measured with and without the material between the plates with $K = C_s-C_p/C_a$ where C_s = total capacitance, C_p = residual capacitance, C_s-C_p = sample capacitance and C_a = assembly constant.

Dissipation factor of a sample is the power loss across the material. This is calculated as follows $D = (C_s/C_s-C_p)$ Dbr x f where D_s = total dissipation factor, D_p = residual dissipation factor, Dbr = D_s - D_p and f = frequency.

The bulk resistivity at different frequencies is calculated from the relationship $P_f = (1/2\pi f\ C_pD)(10^{12})(A/t)$ and should approach the bulk resistivity value obtained employing DC measurements.

In an attempt to remove surface water, the pellets were added to a dry nitrogen flushed desiccator, evacuated and reexposed to atmospheric pressure by application of dry nitrogen and transferred to the sample cell.

DC measurements were made on electrically aged samples, i.e., the voltage was applied for about 5 minutes or until electrical stability was obtained before measurements were made.

DISCUSSION AND RESULTS

We have reported DC bulk resistivity values for a number of metal containing polymers. Those containing Group IV A,B metals and uranium were typically semiconductors with bulk resistivities in the range of 10^3 to 10^{10} ohm-cm [3,4]. Polymers derived from Group V A elements were typically near semiconductors with resistivities in the 10^{10} to 10^{12} ohm-cm region except where whole-chain resonance was possible where bulk resistivity decreased to 10^5 to 10^9 ohm-cm.

Electrical measurements are sensitive to factors as moisture, sample preparation, impurities, etc. Values are reproducible from sample to sample to ± 10% and less as pellet thickness is varied from about 0.1 to 1.0 cm indicating a lack of importance of surface contributions.

All of the previously reported values were obtained employing DC measurements. Correlation with AC measurements would be meaningful in determining if in fact these products maintain semiconductivity within an AC field. Table 1 contains representative values for dielectric constant (K), dissipation factor (D) and bulk resistivity (p) and nature of the metal.

Values for dielectric constant (60 Hertz) for typical polymers vary from about 2.0 for poly(tetrafluorethylene) to a high around 8.4 for poly(vinylidene fluoride) with values for nylons (3-4.5), poly(carbonates)[3], polyesters[3,4] typically being in the range of 3-5, indicative of products possessing relatively (electrically) viscous moieties[5]. Dissipation factors (60 Hertz) range from about 10^{-4} for poly(styrene) to 0.12 for poly(vinyl butyral) with typical condensation polymers exhibiting values from about 10^{-3} to 4 x 10^{-2}[5].

Values of D for the polyoximes typically range from 2.6 to 9.8 (at 10^3 Hertz) and for K from 5 x 10^{-2} to 0.58 (at 10^3 Hertz) and are within the medium to high range of values found for typical polymers (Table 1).

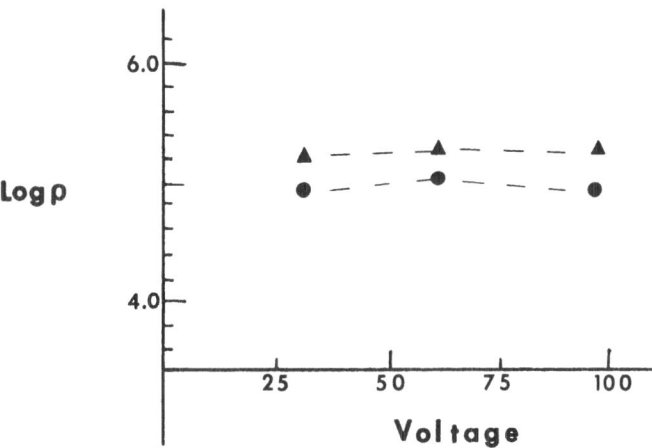

Fig. 1. Bulk resistivities for polyoximes from vitamin K_3 and Cp_2TiCl_2 ● and Cp_2ZrCl_2 ▲ at 500 lbs/square inch applied pressure

Table 1.   Electrical Properties as a Function of Polyoxime.

Electrical Property	Po[a]	1			5			10		
Frequency (10^3)		K	D	$P_{1,000}$	K	D	$P_{5,000}$	K	D	$P_{10,000}$
Compound										
Ti-benzoquinone dioxime	1×10^4	8.8	0.25	1.3×10^9	7.4	0.14	7.3×10^7	6.8	0.098	5.2×10^7
Zr-benzoquinone dioxime	8×10^4	4.9	0.58	6.2×10^7	2.8	0.53	1.3×10^7	1.8	0.86	4.1×10^8
Hf-benzoquinone dioxime	4×10^6	2.6	0.19	3.4×10^8	2.1	0.12	1.1×10^8	1.8	0.014	4.6×10^8
Ti-Vitamin K_3	9×10^5	9.3	0.4	1.0×10^8	6.8	0.13	6.4×10^7	7.2		
Zr-Vitamin K_3	2×10^6	5.9	0.19	3.7×10^8	5.1	0.12	1.2×10^8	4.8	0.080	8.7×10^7
Hf-Vitamin K_3		9.8	0.37	2.1×10^8	7.2	0.21	7.4×10^7	6.5	0.13	6.0×10^7
Zr-1,4-cyclohexanedione dioxime		5.1	0.05	8.7×10^8				4.8		

a.   AT 30 VOLTS APPLIED-DC., 5000 lbs/in^2 APPLIED PRESSURE

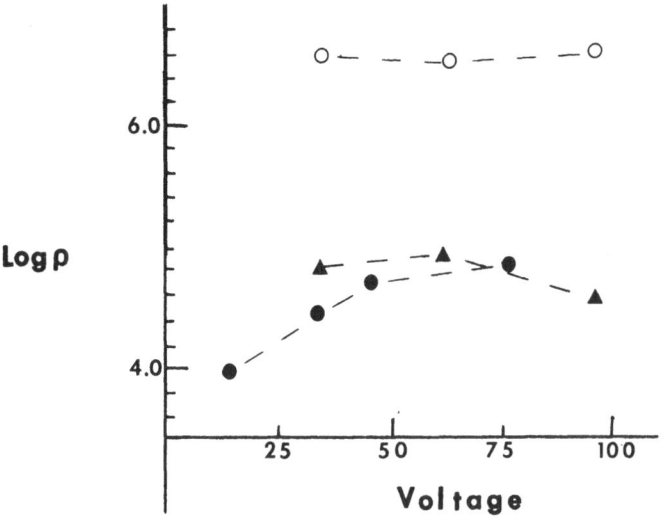

Fig. 2   Bulk resistivities for polyoximes from benzoquinone
          dioxime   and Cp_2TiCl_2 ●, Cp_2ZrCl_2 ▲ and Cp_2HfCl_2 o
          at 5000 lbs/square inch applied pressure.

    The K and D values indicate a generally highly electrically
viscous material consistent with highly polar units of a semiconductor.   This is reinforced by the partial inner correlation between K and D where high K values and high D values are found for
the same polymers.

    Bulk resistivity, DC values, are in the range of 10^4 to 10^6 ohm
cm, within the lower range for conductors.   The relationship between the AC (10^7 to 10^9 ohm cm at 1 KHz) and DC resistivities
indicates that the major dipole contributions to the electrical
properties have not yet occurred to a frequency of 1 KHz .

    Figures 1 and 2 show bulk resistivities as a function of Group
IV B metal.   For the polyoximes derived from the vitamin K_3 and
benzoquinone dioxime, bulk resistivity remains approximately constant from a voltage of 30 to 100, indicative that the number of
current (charge) carriers is approximately constant over this voltage range and that the charge carriers are either intricately

inherent in the polymer nature or one formed at lower voltages.

The effect of chain length on the various electrical parameters was studied using a series of polyoximes of varying lengths.
The polyoximes were synthesized using cyclohexanone oxime as the
chain terminator. Thus menadioxime containing 1 to 40 mole-% of
cyclohexanone oxime was condensed with Cp_2TiCl_2 resulting in chain
lengths varying from about 250 to 3. It can be assumed that the
majority of chains are capped with end groups derived from cyclohexanone oxime minimizing the importance of endgroups. Results
appear as figures 3 and 4. The bulk resistivities for both the
products prepared from 1,4-cyclohexanedione dioxime and benzoquinone dioxime are about 10^4 to 10^5 ohm cm and for each series the
values are within a decade of one another indicative that chain
length is not a critical factor in determining the bulk resistivity. The bulk resistivity for the monomeric model compound derived from condensation of Cp_2TiCl_2 with cyclohexanone oxime exhibits a considerably higher bulk resistivity (ca. 10^8 ohm cm), but
this may be a consequence of the R-M-R, R-M-R, R-M-R interactions
compared to those present in the oligomeric and polymer chains
where the internal repeat unit is -R-M-R-M-R-M- rather than a direct function of chain length.

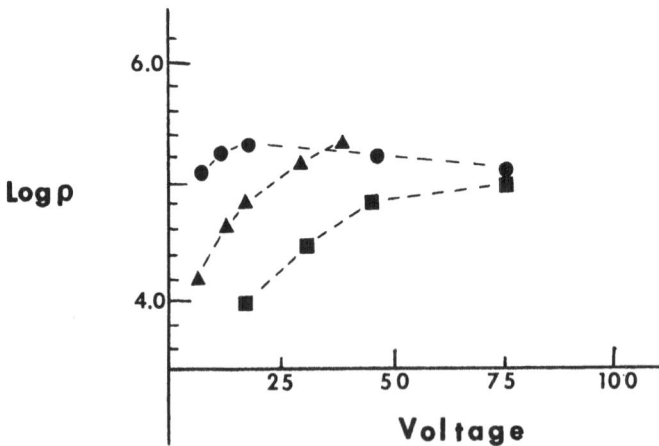

Fig. 3.   Bulk resistivities for polyoximes from benzoquinone
          dioxime and Cp_2TiCl_2 employing cyclohexanone oxime
          as the chain terminating agent where chain length is
          ■ > ● > ▲ at 5000 lbs/square inch applied pres-
          sure.

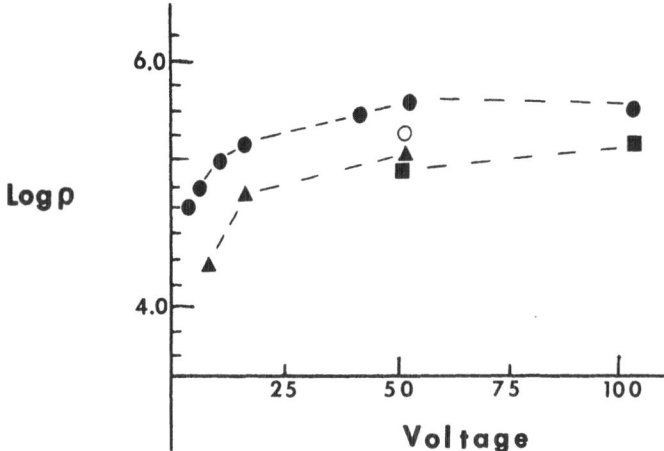

Fig. 4   Bulk resistivities for polyoximes from 1,4-cyclohexanedione
dioxime   and Cp_2TiCl_2 employing   cyclohexanone   oxime
as the chain terminating agent where chain length is
■ > ▲ > ○ > ● at 500 lbs/square inch applied pressure.

Whole chain resonance, where a charge can be pictured through
resonance considerations to flow throughout the chain, has been an
important factor in some other organometallic polymer systems. For
instance, the polyoxime derived from triphenylantimony dichloride
and condensation with benzoquinone dioxime exhibits a resistivity
of about 10^6 ohm cm whereas other analogous products incapable
of exhibiting whole chain resonance show resistivities typically
above $10^{10\ 3}$. For the present system there appears to be little
difference between the polyoximes capable of exhibiting whole chain
resonance (such as those derived from the dioximes of benzoquinone
and vitamin K_3) and those derived from cyclohexanedione dioxime.

Of interest is the physical structure of the polyoximes. The
Group IV B oximes all form fibers on mechanical agitation. The
fibers can be a centimeter in length and 10^{-5} to 10^{-7} centimeters in
diameter. They are flexible with some dimensional strength. If the
single fibers offer equal or better electrical conductivity, many
applications can be envisioned for such conductive fibers. It must
be noted that thus far measurements have been taken only on the
materials in the bulk.

REFERENCES

1. C. Carraher and L. Torre, Organic Coatings and Plastics 18 (1980).
2. C. Carraher and L. Torre, unpublished results.
3. C. Carraher, J. Schroeder, W. Venable and C. McNeely, Organic Coatings and Plastics Chemistry, $\underline{38}$, 544 (1978).
4. C. Carraher, D. Leahy and S. Ailts, Organic Coatings and Plastics Chemistry, $\underline{37}$, 201 (1977).
5. J. Brandrup and E.H. Immergut (Editors), "Polymer Handbook," 2nd Edition, John Wiley & Sons, N.Y., 1975 (pp VIII-7,8).
6. E.J. Murphy and S.O. Morgan, Bell System Tech. J., $\underline{16}$, 493 (1937).

ELECTRIC CONDUCTIVITIES OF POLY(VINYL KETONE)S
REACTED WITH PHOSPHORYL CHLORIDE

Takeshi Ogawa, Ruben Cedeño, Tito E. Herrera,

Benjamin Almaráz, and Motomichi Inoue

Centro de Investigación sobre Polímero y Materiales
Escuela de Ingenieria, Universidad de Sonora
Apartado Postal A-136, Hermosillo, Senora, México

INTRODUCTION

Previously[1] we have reported the new reaction of poly(alkyl
vinyl ketone)s with active chlorides such as phosphoryl chloride,
to obtain the corresponding poly(acyl actylene)s

and later the changes in the electric conductivities of poly(ethyl
vinyl ketone) films during the reaction with phosphoryl chloride
were reported.[2] The poly(ethyl vinyl ketone) film showed its specific conductivity of up to 10^{-2} Ω^{-1} cm^{-1} when reacted at 0°C after
about 200 hours of reaction in petroleum ether. In this paper the
electric conductivities of poly(methyl vinyl ketone), poly(ethyl
vinyl ketone) and poly(phenyl vinyl ketone), which were reacted with
phosphoryl chloride and thiophosphoryl chloride in solution and in
the form of a film in non-solvent, are reported.

EXPERIMENTAL

(1) Materials

Methyl vinyl ketone (MVK) used was a commercial product and it
was distilled in vacuum before polymerization. The polymerization
was carried out at 50°C in benzene (50 % v/v) using 2,2'-azobisiso-
butyronitrile as an initiator for the low molecular weigh polymer,
and at room temperature without initiator for the high molecular

weight polymer. Ethyl vinyl ketone (EVK) was prepared according to
the method reported by McMahon et al[3], and it was purified by disti-
lling twice under a reduced pressure before the polymerization at
50°C in bulk using the same initiator as above. Phenyl vinyl ketone
(PVK) was prepared by the method reported by Allen et al[4]. The distilled monomer polymerized on standing at room temperature, and this
polymer was used for the reaction. All of these polymers were purified by reprecipitation from chloroform-methanol system. They had
the following intrinsic viscosities:

      Poly-MVK-1 : 0.70 in chloroform at 30°C
      Poly-MVK-2 : 3.34 in chloroform at 30°C
      Poly-EVK   : 0.50 in chloroform at 30°C
      Poly-PVK   : 0.45 in Benzene at 30°C.

Phosphoryl chloride and thiophosphoryl chloride were distilled once
before use, and tetrahydrofuran was distilled over lithium alminium
hydride. Chloroform was distilled over phosphorus pentoxide. Hexane
and petroleum ether were purified by a single distillation.

(2) Reactions

 i) In solution: 1 g. of poly-MVK was dissolved in 100 ml of chloroform or tetrahydrofuran and 5 ml of phosphoryl chloride were added
and the mixture was kept in an ice bath for a required time. After
the reaction a required amount of triethylamine and excess of methanol were added at 0°C and the system was concentrated under vacuum
and the reacted polymer was precipitated in petroleum ether, washed
rapidly with methanol and dried in vacuum. The reactions of poly-
EVK and poly-PVK were carried out similarly but in the case of poly-
PVK nitrobenzene and xylene were used as solvents because poly-PVK
did not react with reasonable rate at temperatures below 100°C.

ii) In the form of film: The films of poly-MVK and poly-EVK were
prepared by casting the chloroform solutions of these polymers on a
mercury surface, and they were dried at room temperature for a few
weeks. Films with thicknesses of about 0.2 - 0.7 mm were prepared.

(3) Conductivity measurements
     For the powder samples, a pressure mold shown in Fig. 1 was
used. The powder was compressed under a nitrogen atmosphere to a
disk 5 mm diameter and 0.5 mm thickness. For the films a glass
made apparatus shown in Fig 2 was used. A small piece of polymer
film, 5 x 8 mm, was connected to thin gold tapes which were connected
to tungsten wires (Fig. 2 b). For the films of poly-MVK-2, which is
flexible and strong, they were connected directly to the tungsten
wires (Fig. 2 c). In both cases a carbon black paste was used for
the connecting parts. A V-shape apparatus was used previously[2] for
monitoring, but it was found not necessary and the single tube apparatus was used in this work (Fig. 2 a). The measurements were
carried out using alternating current of less than 20 Hz.

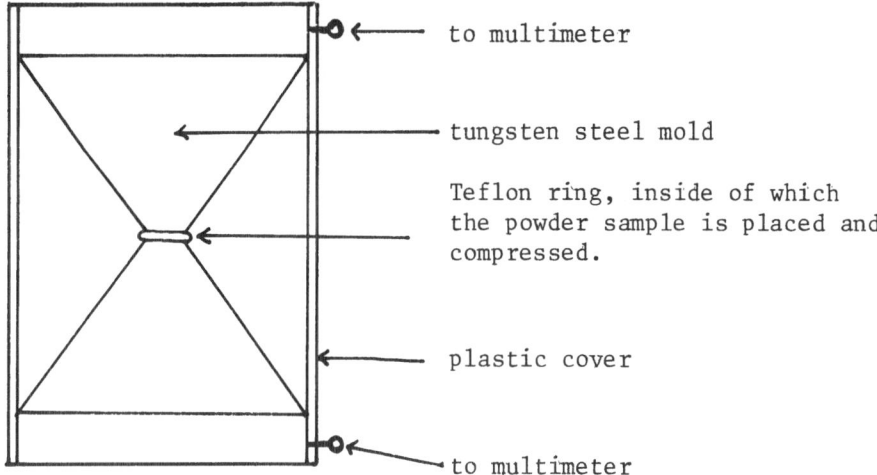

Fig. 1. The apparatus for conductivity measurement for powder samples.

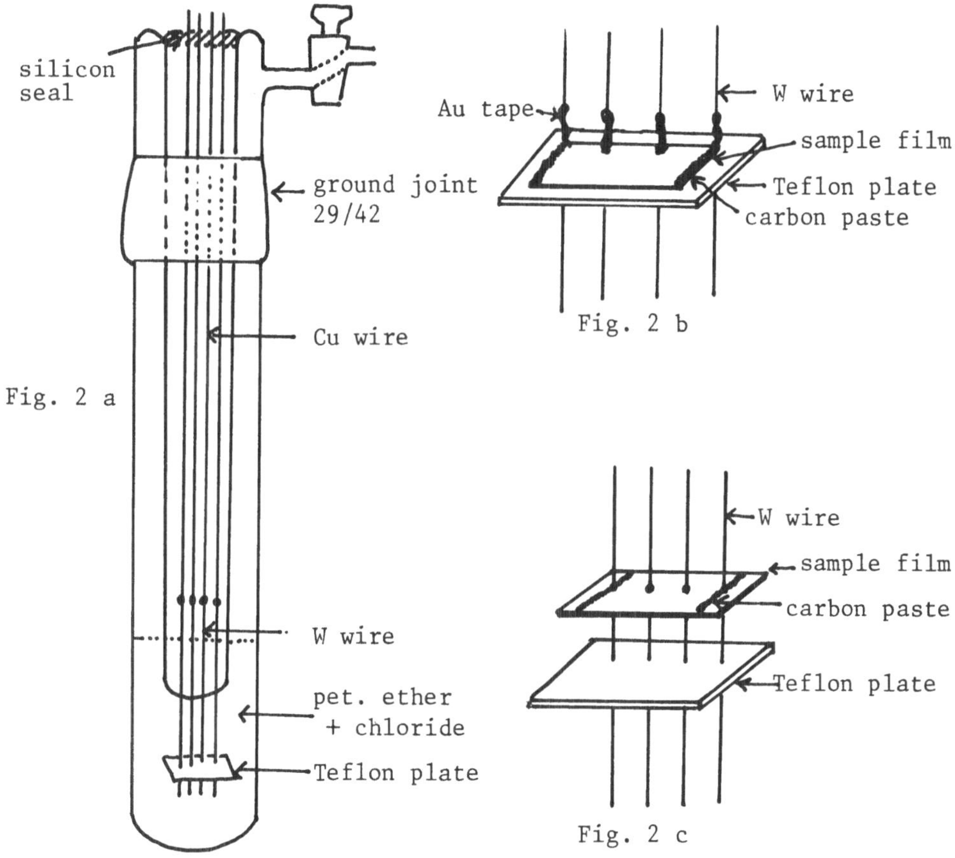

Fig. 2. The apparatus for the reaction and conductivity measurement
for polymer films.

RESULTS AND DISCUSSION

The relationships between the apparent conductivity of low molecular weight poly-MVK-1 films and the reaction time are shown in
Fig.3. As can be seen from the figure the maximum conductivity is not
very consistent between the samples, some having values near 10^{-2}
$\Omega^{-1}cm^{-1}$ and others having only 10^{-4} $\Omega^{-1}cm^{-1}$. Furthermore the second
increase in the conductivity (curve C), which was observed for most
cases of poly-EVK[2], did not appear for some samples in the case of
poly-MVK. The reason for this irregularity is not known at present
but it is thought that the reaction temperature of 0°C is still too
high for the poly-MVK-1, which apparently completes the reaction in
30 hours whilst poly-EVK takes about 200 hours to complete the reaction. Studies at temperatures below 0°C are being made at present.

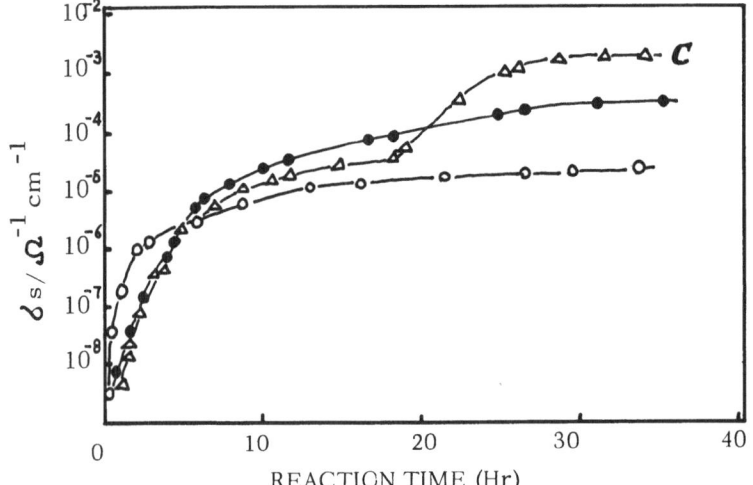

Fig. 3. Changes in the conductivity with reaction time.
Poly-MVK-1 film. Reacted at 0°C. in pet. ether.
$(O=PCl_3)$ = 4 % v/v.

The films of poly-MVK-1 after the reaction were no longer soluble
in solvents, indicating that the acid catalyzed condensation took
place. This may be the cause of the conductivity being poorer than
in the case of poly-EVK films, which remained soluble after the reaction at 0 °C. As in the case of poly-EVK[2], when the reacted
poly-MVK-1 films were washed with pet. ether and dried in vacuum at
40°C for 24 hours, the conductivity decreased to about 10^{-5} Ω^{-1}
cm^{-1} when measured at 0°C. The activation energy of the conductivity for the reacted poly-MVK-1 films was found to be 0.5 - 0.8 eV
over the temperature range of 0 - 60°C.

The powder samples of poly-MVK-1 which precipitated after the reaction with phosphoryl chloride at 0°C in solution (THF or chloroform) for 7 days, were filtered, washed with methanol containing triethylamine and dried under vacuum. They showed a conductivity of the order of 10^{-5} Ω^{-1} cm^{-1}, when measured in the form of disc using the apparatus shown in Fig. 1.

The apparent specific conductivity of poly-EVK films during the reaction depended on the reaction temperature[2]. The lower the temperature the higher the electric conductivity of the product. In this work a change in the reaction temperature was made during the reaction, and the result is shown in Fig. 4.

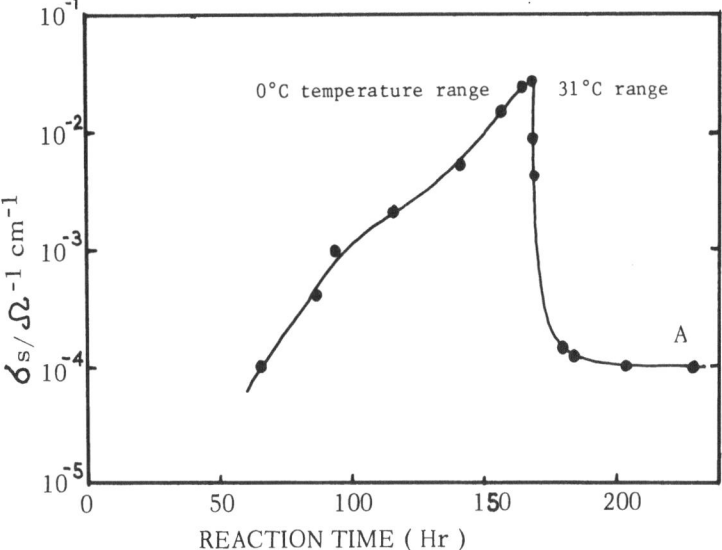

Fig. 4. Temperature effect on the conductivity of poly-EVK film. $(O=PCl_3)$ = 4 % in hexane.

It can be seen that the conductivity of the film decreased drastically with the temperature increase, contrary to the increase in the conductivity that would be expected due to the activation energy of about 0.5 eV. The conductivity remained constant for a period of 50 hours after raising the temperature. When the temperature of the system was decreased again to 0°C at the point A in Fig.4, the conductivity decreasd to 7 x 10^{-5} Ω^{-1} cm^{-1} from 10^{-4} Ω^{-1} cm^{-1}. This corresponds to the change due to the activation energy of δ_s.

The decrease in the conductivity when the temperature was raised from 0°C to 45°C seems to be a chemical effect. If the acid-cata-

lyzed Aldol type condensation between the carbonyl and methyl groups
had taken place when the temperature was raised, the conductivity
shoul have increased because of the resulting increase in the passage of electric current. The same poly-EVK was reacted with phosphoryl chloride in a mixed solution of chloroform and tetrahydrofuran
ay 0°C and 45°C for 7 days. The reacted polymers were precipitated
in methanol containing triethylamine, washed well with methanol
and dried under vacuum. The polymers were analyzed for their chlorine and phosphorus contents. Although there remains the problem
whether the washing of the polymers was complete or not, the poly-
EVK reacted at 0°C contained 0.79% of Cl (corresponding to 1.6 atoms
per 100 MVK units) and less than 0.1% of P (corresponding to 2.3
atoms per 1,000 MVK units), while the poly-EVK reacted at 45°C contained 1.3 % of Cl (2.7 atoms per 100 MVK units) and 0.25 % of P
(5.7 atoms per 1,000 MVK units). From these results it is thought
that the addition of HCl takes place more readily at higher temperatures, which decrease the degree of conjugation.

    Similar experiments to above were carried out for poly-MVK-2,
and the result is shown in Fig. 5.

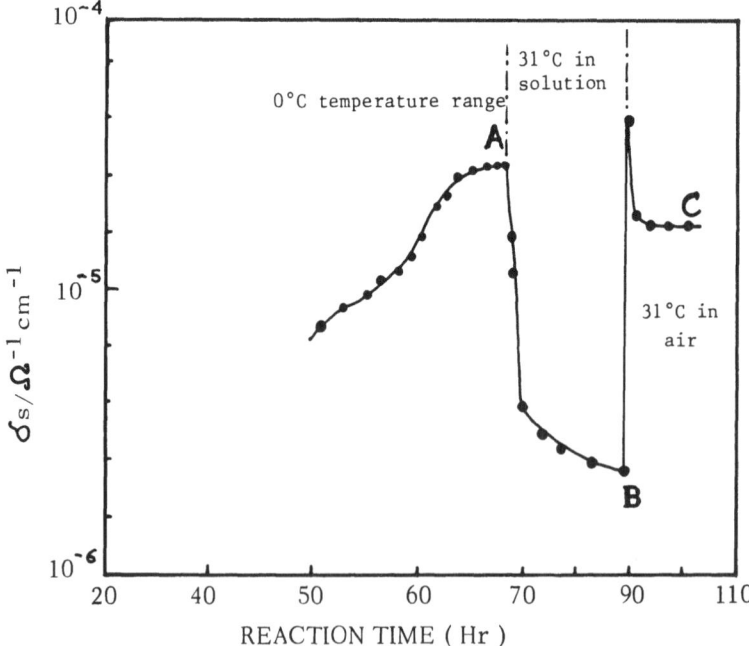

Fig. 5. Conductivity change with reaction time for poly-MVK-2 film.
        (O=PCl$_3$) = 7 %. Film thickness = 0.635 mm.

As can be seen from the figure, the reaction of the poly-MVK-2 film
was very slow compared to the case of that of poly-MVK-1. It is
thought that the higher the molecular weight of polymer, the slower
the penetration of phosphoryl chloride. The maximum conductivity
at 0°C in the figure was not as high as that of the case of poly-
MVK-1, but further studies are being made for poly-MVK-2 varying the
reaction conditions. It can be seen that, when the reaction temperature was raised to 31°C (point A), the conductivity decreased similarly to the case of poly-EVK (Fig.4). When the sample was exposed
to air, i.e. taken out from the solution, the conductivity increased
instantaneously, but soon became constant again. The final product
(point C) was kept in atmosphere for several weeks, and it maintained
its conductivity of the order of 10^{-5} Ω^{-1} cm^{-1}.

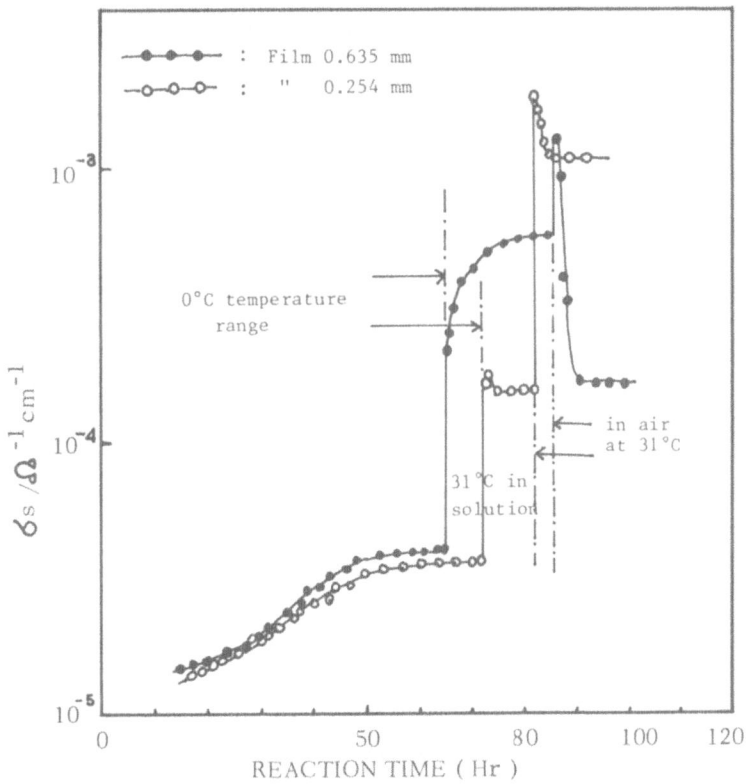

Fig. 6. Changes in conductivity with reaction time for poly-MVK-2
        films reacted with S=PCl$_3$ in pet. ether.

Fig. 6 shows the same reaction of poly-MVK-2 in pet. ether using thiophosphoryl chloride. The conductivity at the temperature range of 0°C, was similar to that in the case of phosphoryl chloride, being of the order of $5 - 6$ x 10^{-5} $\Omega^{-1}cm^{-1}$. However when the reaction temperature was raised, the conductivity also increased drastically, this being completely opposite to the case with phosphoryl chloride. The reason for this difference in the actions between phosphoryl chloride and thiophosphoryl chloride, is not known yet, and detailed studies are being made at the present. It is thought tentatively however, that the thiopfosphoryl group is less polar than the phosphoryl group $(O=P- \rightleftharpoons {}^-O-P^+)$, and therefore the phosphoryl group could interact with the conjugated system having low electronic density due to the carbonyl groups, facilitaing the addition of HCl at higher temperatures, whilst the thiophosphoryl group would not interact with the conjugated system.

The thickness of the film did not cause appreciable difference in the conductivity at the region of 0°C, but the conductivities after raising temperature and exposing to air were different depending on the film thickness. The changes in the conductivity when exposed to atmosphere is the problem of absorption of water. The results on the effects of doping with various substances will be reported in future.

Fig. 7 shows the same reaction, but at 31°C instead of 0°C. The reaction was much faster, and the conductivity reached a maximum in about 5 hours and it started to decrease slowly. This phenomenon of having a optimum reaction time for the maximum conductivity was also observed in the case of poly-EVK films reacted with phosphoryl chloride at 30 and 40°C[2]. When the film was exposed to atmosphere the conductivity increased but rapidly decreased again and became more or less constant, being $7 - 9$ x 10^{-5} $\Omega^{-1}cm^{-1}$, for the experimental period of 60 hours.

It is noteworthy that all of these poly-MVK-2 films reacted with these chlorides were intensely black with shinning surface, and they do not change their appearence even when kept in air for several weeks and do not lose their conductivities. And they maintain the strength and flexibility.

Poly-PVK does not react at low temperature such as the boiling temperatures of chloroform, benzene, etc. Only when solvents with high boiling points such as toluene, xylene, nitrobenzene, etc. were used, a black powder was obtained by the reaction with phosphoryl chloride at temperatures above 100°C. The reason for this is thought to be steric effect of phenyl groups on the formation of conjugated sp^2 units. Because of the high temperature of reaction the conductivity of the reaction product is expected to be poor. Furthermore the steric effect will not allow the formation of highly conjugated system, and at high temperatures such as above 100°C various side reactions, including the main chain scission will take place when reacted with active reagents such as phosphoryl chloride.

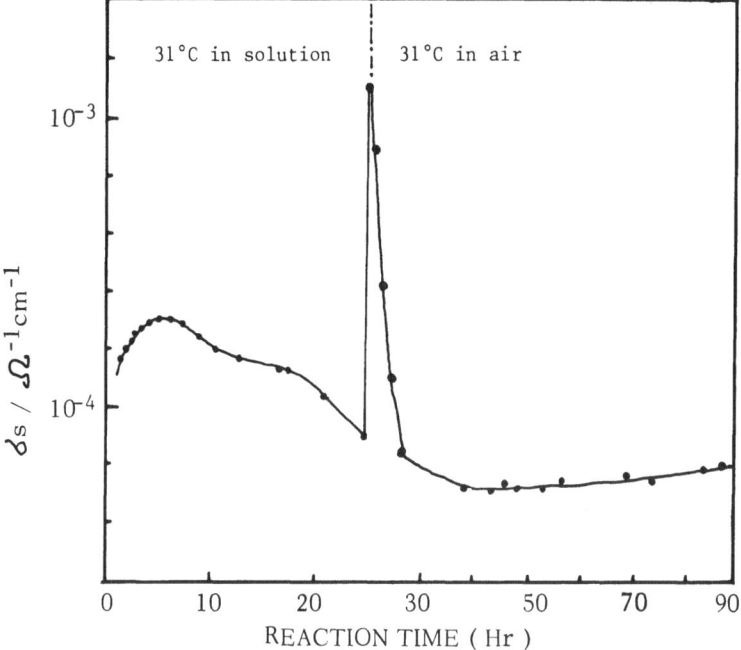

Fig. 7. Change in the conductivity of poly-MVK-2 film reacted with
thiophosphoryl chloride at 31°C. (S=PCl₃) = 7 % v/v.

Table 1. Apparent conductivities of poly-PVK reacted with phosphoryl
chloride at 100°C.

Sample	solvent	reaction time	measured in	conductivity ($\Omega^{-1}cm^{-1}$)
A-1	nitrobenzene	24 hours		negligeble
A-2	xylen	24 "	in air	3×10^{-5}
A-4*	"	24 "	in air	4×10^{-5}
A-4*	"	24 "	in vacuum	1×10^{-6}
A-5**	"	24 "	in air	4×10^{-4}
A-5**	"	24 "	in vacuum	1×10^{-7}

A-1 and A-2 : 1 g of poly-PVK, 6 ml of O=PCl₃, 20 ml of solvent.
A-4 and A-5 : 0.7 g of poly-PVK, 4 ml of O=PCl₃, 20 ml of solvent.
* Insoluble part, ** Soluble part after the reaction.

        Fig. 8 shows NMR spectra of poly-PVK and its reaction products.
It can be seen that the majority of the peaks due to the methylene
and methyne protons disappeared after the reaction. The apparent
conductivities of poly-PVK reacted with phosphoryl chloride, measured

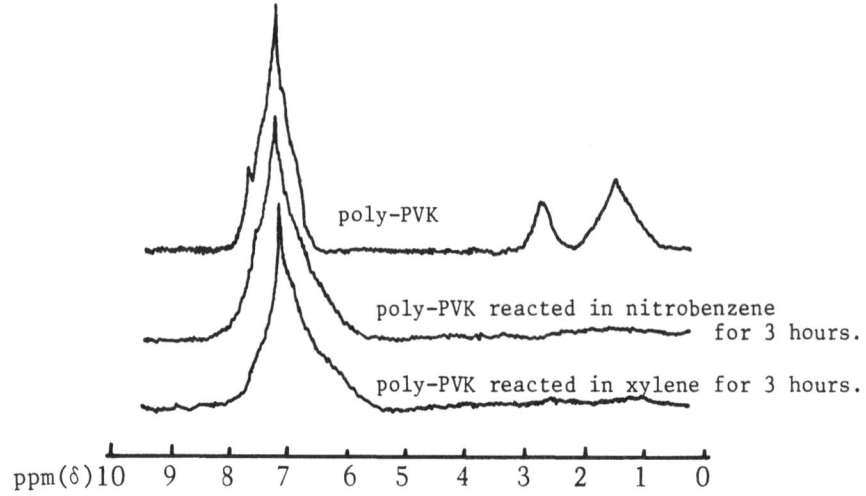

poly-PVK

poly-PVK reacted in nitrobenzene
for 3 hours.

poly-PVK reacted in xylene for 3 hours.

ppm(δ)10 9   8   7   6   5   4   3   2   1   0

Fig. 8. NMR spectra of poly-PVK and its reaction products.

in the form of disc, are shown in Table 1. As can be seen from the
table, the conductivities in vacuum are very low, and it seems that
water molecules are absorbed when exposed to atmosphere, which increase the conductivity. It may be said that poly-PVK is not suitable for obtaining a semiconductive polymer by the method of this
work.

REFERENCES

1. T.Ogawa, R.Cedeño and T.E.Herrera; Makromol. Chem., 180, 785
   (1979)
2. T.Ogawa, R.Cedeño and M.Inoue; Polymer Bulletin, 2,275 (1980)
3. E.M.McMahon, J.N.Roper Jr, R.H.Harris and R.C.Brant;
   J.Am.Chem.Soc., 70, 2971 (1958)
4. C.H.Allen and W.E.Baker; J.Am.Chem.Soc., 54, 740 (1932)

UNUSUAL CHARGE TRANSFER AT A PHOTOCONDUCTOR-COPOLYMER INTERFACE:

ITS ROLE IN A NOVEL PARTICLE MIGRATION MICROFILM

Jack Y. Josefowicz and Chung C. Yang

Xerox Research Centre of Canada
2480 Dunwin Dr.
Mississauga, Ontario, Canada L5L 1J9

INTRODUCTION

Studies of the charge-transport mechanism between a photoconductor and a polymer are of both practical and fundamental interest. Such a mechanism is intimately involved in a particle migration imaging microfilm first described by Goffe[1,2]. This film consists of a monolayer of Se spheres located near the surface of a thermoplastic matrix which is supported in a conductive transparent substrate as shown by the electron micrograph in Fig. 1. Sensitization of the film is accomplished by electrically charging the surface of the thermoplastic with a corona discharge while the conductive substrate is connected to ground. Exposure of the film causes photogeneration of electron-hole pairs in the Se spheres which separate under the influence of the applied electric field. During this process charge transport occurs within the Se-thermoplastic system with the end result that the Se spheres attain a net charge. The charged image can be developed by heating the film above the glass transition of the polymer matrix; as the viscosity of the polymer decreases the negatively charged Se spheres migrate towards the Al electrode under the influence of their image electrostatic force. The Se spheres which are not exposed to light, remain neutral and do not migrate. Upon exposure of the film to solvent, the surface of the film containing the neutral Se spheres can be removed. Studies of the optical properties[3,4,5] and electrical properties[6], as well as the photographic sensitivity[6], have recently been reported. However, the details of the microscopic charge transport, which is intimately linked to both the imaging and development processes in this film still required clarification. In this paper, we present a study which elucidates the details of the charge transport mechanism during the imaging process in a

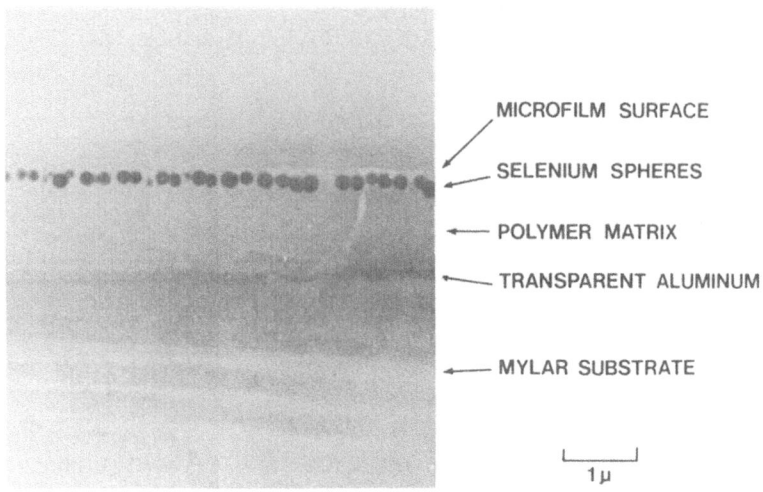

Fig. 1: Thin section electron micrograph of a particle migration
microfilm. The average Se sphere diameter is 0.3um and the
thickness of the styrene hexylmethacrylate polymer matrix
is 1.5um.

particle migration microfilm. This was accomplished by investi-
gating thin film structures composed of the microfilm sandwiched
between semi-transparent Al electrodes; as well as continuous
thin film structures composed of Se and styrene-hexylmethacrylate
which were also sandwiched between Al electrodes. Charge transport
within these thin film structures was studied using the technique
of photostimulated current transients. The results of this study
have led to the uncovering of an usual rate-limited charge-transfer
mechanism across the Se-copolymer interface; which occurs during
the imaging process in the particle migration microfilm.

The first portion of this paper deals with the investigation

of charge transport processes within the particle migration microfilm. In the latter part of the paper, using sandwich cells composed of continuous thin films of Se and the styrene–hexylmethacrylate copolymer, the details of the charge transfer mechanism at the
photoconductor–polymer interface are investigated.

SAMPLE PREPARATION

Electroded microfilm sandwich cells as shown by the structure
in Fig. 2, were prepared as follows. A sharp edge was produced in
the end portion of the microfilm by scraping away the thermoplastic

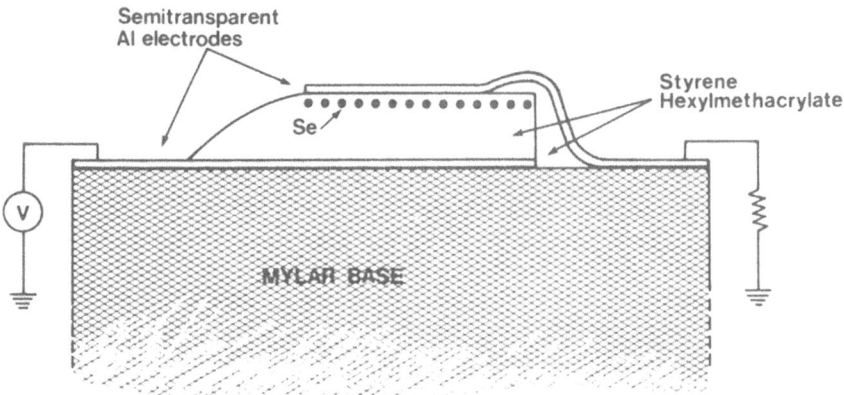

Fig. 2: Electroded microfilm structure. It is composed of an
aluminized (50 A) Mylar Substrate on one side and a vacuum
evaporated semitransparent 150 A film of Al on the surface.
An electric field is applied across the microfilm by
applying a voltage across the two Al electrodes.

as well as the aluminum, leaving the Mylar substrate exposed. The
scraped edge was then coated with a polymer film by passing over it
a Q-tip saturated with a 10% solution of styrene hexylmethacrylate
copolymer[7] in toluene. This was followed by vacuum deposition of
Al (150 Å) at 25^0C over top of the microfilm, thereby creating an
electroded sandwich structure across which an electric field could
be applied. Additional sandwich structures of the type, Al-copolymer-
Al, were made by dip coating ultra thin copolymer films (200 A -
3000 Å) on the same aluminized Mylar substrate that was used in
making the particle migration microfilm and then vacuum evaporating
a   150 Å Al semitransparent film. Coatings and evaporations were
performed at 25^0C.

    The continuous thin film sandwich cells were prepared on the
same aluminized Mylar substrates as was used in the microfilm[8]. The
polymer was dip coated[9] onto the aluminum from solutions of pure
toluene (Eastman Kodak, Grade Spectral acs). The thickness of the
polymer coating on the substrate could be varied between 100 Å and
1μm. This was verified by observing thin sections of these films
with the electron microscope. After being coated the polymer films
were placed under high vacuum (10^{-7} Torr) for 2h at 20^0C. This was
followed by the evaporation of a 500 Å to a 1000 Å layer film of Se
onto the polymer coating, which was at a temperature of 25^0C. To
complete the sandwich cell a 150 Å film of Al was evaporated over
the Se. The capacitance of the sandwich cells was measured with a
Hewlett-Packard 4265B capacitance bridge. A diagramatic representation of the thin film structure is shown in Fig. 3a.

EXPERIMENTAL METHOD

    The method used to investigate the charge transport phenomena
in the thin film sandwich structures was photo-stimulated current
transients. A schemative representation of the experimental set-up
is shown in Fig. 4a. A constant voltage power supply is connected
in series with the electroded sandwich cell which is in series with
a load resistor of 100 ohms. A typical time constant of the circuit
is approximately 1 usec. Induced current in the external circuit
due to charge transport within the electroded film produces a
voltage drop across the load resistor which is amplified and recorded
by a Biomation transient recorder. The data can be stored temporarily
in a signal averager or recorded on magnetic tape. The data on tape
is sent to a computer for analysis and a hard copy of the results
can be produced. For the light source there was a choice of either a
Xenon flashlamp, which has a pulse width of 2.5 usec and a pulse
intensity of 10^{14} photons/cm^2, or a pulsed dye laser with a pulse
width of 400 nsec and a pulse intensity of 10^{15} photons/cm^2. A
timing diagram which briefly describes the sequence with which an
experiment is performed is shown in Fig. 4b. A voltage pulse,
50 msec or longer, is applied to the electroded microfilm. Twenty
msec after the voltage pulse is applied, a trigger is sent to the

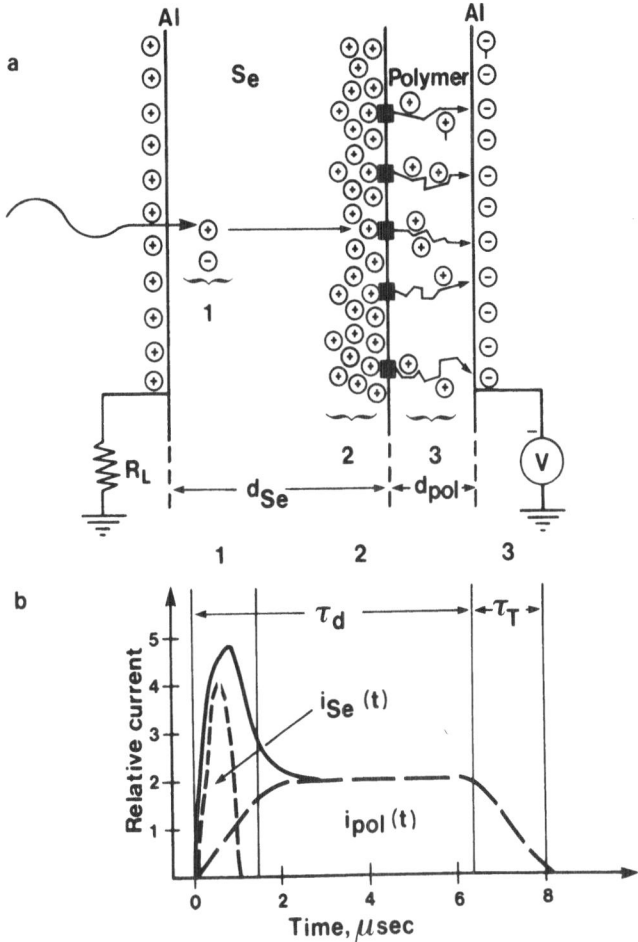

Fig. 3

a) A diagramatic representation of the cross section for the Se-
copolymer sandwich film showing, 1) photogenerated charges which
are separated in Se, 2) rate limited charge transfer through interfacial states, and 3) transport of those charges through the
polymer.

b) A current transient which is generated as a result of the charge
transport processes depicted in part a). Region 1 corresponds to
a current spike associated with charge separation in Se followed
by a decay governed by the time constant of the circuit. Region 2
represents a constant charge-transfer current into the polymer
which persists until the charges in Se are depleted. Region 3
shows a decay in the current immediately after the depletion time
τ_d, which reflects the transit time through the polymer.

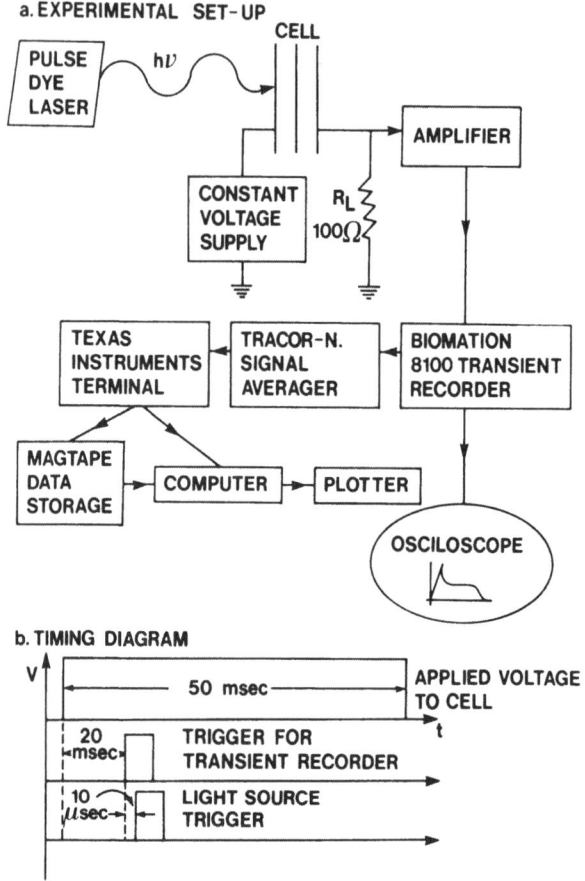

Fig. 4. a) The experimental set-up. A Xenon flashlamp was used as
       the light source which has a pulse width of 2.5usec and an
       intensity of 10^{14} photons/cm^2. The electroded microfilm
       was placed in series with a constant voltage power supply
       and a load resistor across which the voltage was monitored
       by a preamplifier, the output of which was directed to a
       transient recorder.

     b) A simplified timing diagram showing the sequence used in
       generating a photostimulated transient current.

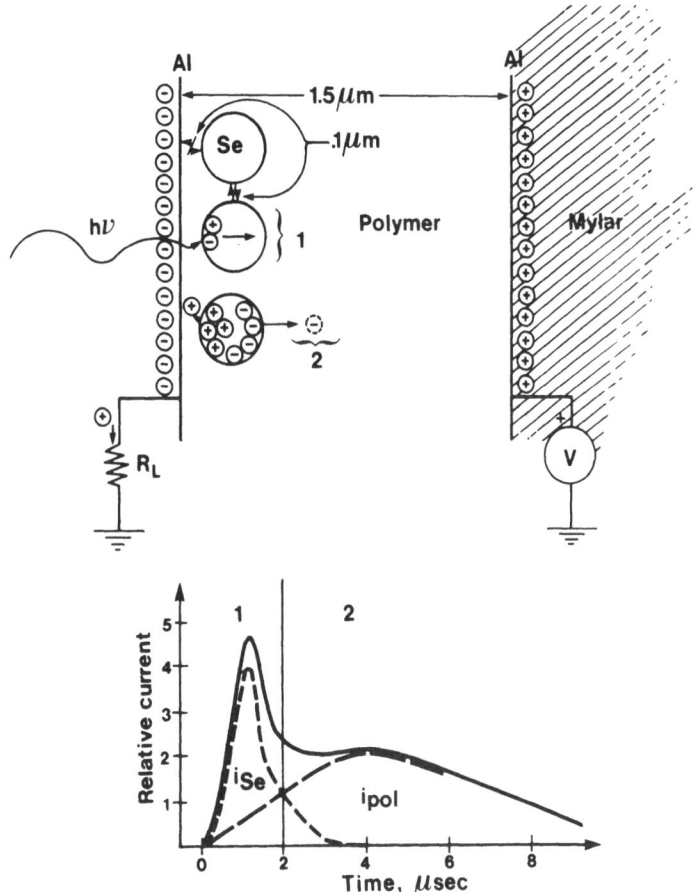

Fig. 5. a) A diagrammatic representation of an electroded microfilm
and associated circuit. Label 1 indicates photogeneration
of electron-hole pairs followed by separation in the
applied electric field. Label 2 represents charge injection across the Se-polymer interface of electrons and
holes.

b) A representation of the expected photo-stimulated current
transient induced in the external circuit due to charge
separation in the Se spheres, labelled i_{Se} and charge
injection across the Se-polymer interface followed by
charge transport in the polymer, labelled i_{pol}.

transient recorder and data acquisition begins. This is followed
10 usec later by a trigger for the pulsed light source.

EXPERIMENTAL RESULTS

Particle Migration Microfilm

    Basic information about the charge transport process in particle
migration microfilm can be obtained from the photostimulated current
transient measurements. The information which can be derived from
such an experiment can be clearly explained using the diagrams
presented in Fig. 5a and 5b. The application of a voltage to the
Al electrode, on the Mylar side of the microfilm, produces an
electric field simultaneously across the polymer as well as the Se
spheres. When a light pulse from the Xenon flashlamp illuminates
the Se spheres, electron-hole pairs are photogenerated which
separate to the right and left extremes of the Se spheres due to
the presence of the applied electric field. This process has been
identified by the label (1) in Fig. 5a. The corresponding induced
external current transient is a current spike which is identified
by the dashed line labelled i_{Se} in portion (1) of the transient
in Fig. 5b. Since the time necessary for charge carriers to transit
the Se sphere is approximately a few nanoseconds, the current spike
associated with this transport represents the integrated current
due to charge separation in Se. This is a consequence of the 1 usec
time constant of the measuring circuit. The width of the current
spike associated with the charge separation process in Se is therefore
a reflection of the intensity profile of the Xenon flashlamp pulse.
If nothing more than charge separation in the Se spheres occurs,
then only the current spike would be detected in the current transient induced in the external circuit. However, if charge injection
occurs across the Se-polymer interface followed by charge transport
in the polymer, it is expected that an additional contribution to
the current transient would be detected at a time beyond the current
spike associated with the charge separation in the Se. This process
is depicted in Fig. 5a by the label (2) and has a corresponding
current transient labelled i_{pol} shown in Fig. 5b. The relatively
longer time scale of the current transient associated with the
charge injection into the polymer is expected since the mobility of
charge carriers in polymers is usually in the range 10^{-6} to 10^{-10}
cm^2/volt-sec range whereas for Se it is orders of magnitude
higher[10,11].

    A superposition of current transient traces on electroded micro-
film for a variation of electric fields between 6.4 x 10^4 V/cm to
1 x 10^6 V/cm across the polymer matrix is shown in Fig. 6. These
results correspond to a positive voltage applied to the Al electrode
on the Mylar side of the microfilm. At relatively low fields of
about 6.7 x 10^4 V/cm, the current transient is characterized by a
spike which reflects the intensity profile of the Xenon light source.

Following the arguments above, it is concluded that this represents charge separation in Se with no subsequent charge transfer into the polymer. When the electric field is increased to approximately 4.0×10^5 V/cm, which approximately corresponds the field at which D_{min} saturates, a second current contribution becomes visible in the tail of the transient; and as the electric field continues to be increased up to 1×10^6 V/cm there is unequibocal evidence of charge transfer from the Se spheres into the polymer. This is clearly indicated by the large current contribution beyond the initial current spike, however, it is not possible to determine from these results whether holes or electrons, or both carriers are transfered across the Se-polymer interface. The details of the charge transfer mechanism, including the charge transfer efficiencies for electrons and holes, are elucidated from the experiments on continuous thin film sandwich structures which are reported in the next section.

In order to ensure that the measured transient current was not associated with the injection of charge from the Al electrode into the styrene hexylmethacrylate copolymer, experiments were performed on Al-polymer-Al sandwich structures, where the polymer thickness varied between 200 Å and 3000 Å. Both in the presence of a Xenon light pulse and in the dark, no measurable current transients were detected for applied electric fields as high as 5×10^6 V/cm.

Another informative feature of the current transients in Fig. 6 is that the relative photocurrent generation efficiency can be determined from the amplitude of the peak heights of the current spike associated with charge separation in the Se sphere. Consequently, the induced current peak amplitude in the external circuit reflects the relative photocurrent generation efficiency. Shown in Fig. 7 are the results of the relative photocurrent peak plotted as a function of applied field for both hole and electron transport in the Se spheres. The expected saturation in photocurrent generation efficiency is reached when the electric field strength is 1.5×10^5 V/cm across the Se spheres. For the case of electrons the relative photocurrent response is less than that for holes until a field of 3.75×10^5 V/cm is reached, where the efficiencies of electrons and holes merge. A similar result has been obtained for thick continuous films of amorphous Se by Hartke and Regensburger[12]. The spectral sensitivity was determined for the Se spheres in the polymer matrix by monitoring the current spike amplitude, at a constant electric field (3×10^5 volts/cm), as a function of wavelength. The result is shown in Fig. 8. There appears to be the expected peak for Se between 400nm and 440nm, after which the film becomes less photoactive and drops off at 540nm. The wavelength dependence is similar to the absorption spectrum of amorphous Se[12].

From both the excitation spectrum and applied electric field dependence of the photocurrent generation efficiency, the

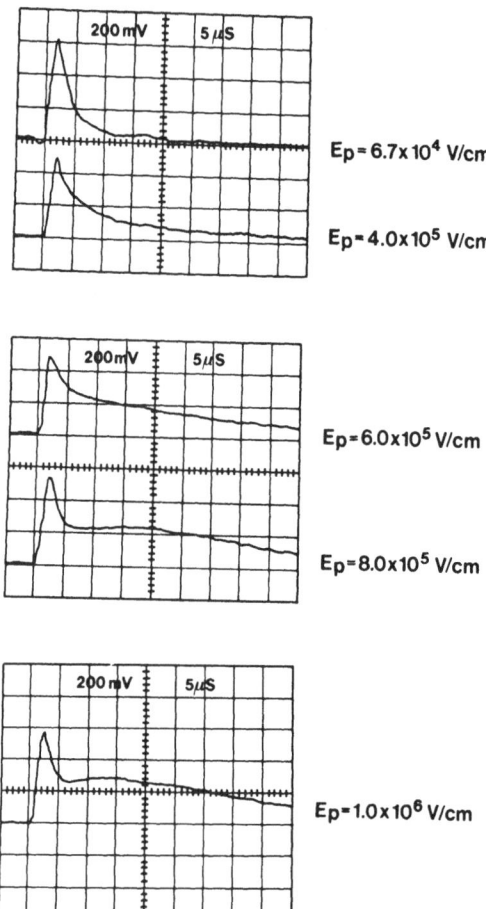

Fig. 6 A series of photostimulated current transient experiments
        where E_p represents the value of the applied electric field
        across the polymer matrix.

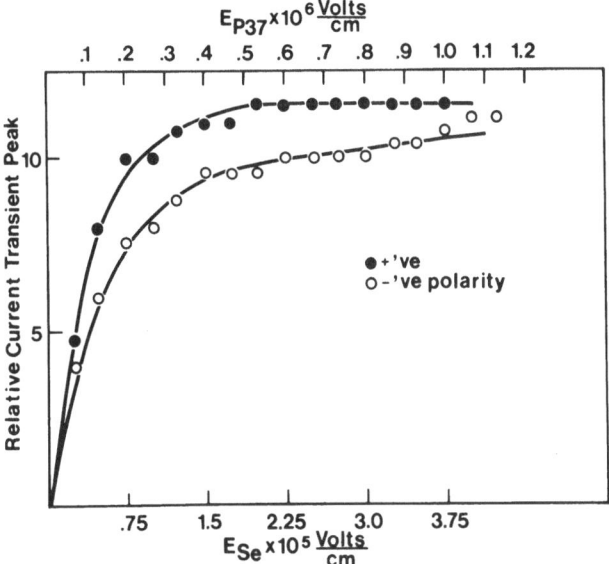

Fig. 7: Electric field dependence of the relative photocurrent
generation efficiency of the Se spheres within the particle
migration microfilm. E_{Se} and E_p represent the values
of the electric field across the Se spheres and the polymer
matrix, respectively.

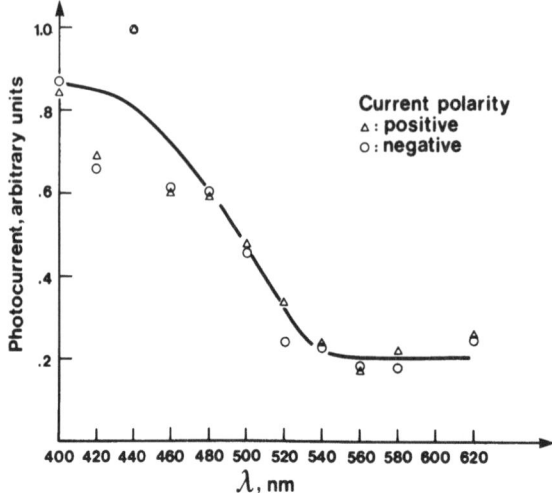

Fig. 8:   The excitation spectrum of the Se spheres within a
          particle migration microfilm.

experimental results show that photosensitivity of the microfilm
generally reflects the photoconductivity properties associated with
Se in the bulk amorphous state.

Continuous Thin Film Sandwich Cells

The results of photostimulated current transient experiments,
performed on the particle migration microfilm, clearly showed that
there is a substantial injection of charge across the interface
between the Se particles and the styrene-hexylmethacrylate copolymer.
In order to distinguish between hole and electron charge transfer
as well as the efficiency with which photogenerated charge in Se
moves across the interface into the polymer matrix, sandwich thin
film structures with a more clearly defined architecture were studied,
as shown in Fig. 3a). With the continuous thin films of Se and
copolymer in this configuration it is possible to distinguish between electron or hole charge transfer across the interface between
Se and the copolymer, depending on the polarity of the constant
voltage power supply.

Recently, photostimulated transient currents have been

measured using a short-duration pulse from a dye laser (10^{15} photon/ cm^2, 0.5 µsec pulse width) which permitted the study of charge-transport phenomena in a time range from approximately 1 µsec to a few seconds[13]. The intensity of the light pulse was large enough to cause complete field collapse in Se for all values of the voltages applied. This approach is convenient in charge-transfer studies, since a well-defined amount of charge is generated in the photoconductor.

Shown in Fig. 9 is a superposition of six photogenerated current transients for a range of electric fields in the polymer from 0.9×10^6 to 5.8×10^8 V/cm, where the sandwich structure was composed of a 700 Å Se film and a 150 Å polymer film. These transients have been scaled in amplitude according to the original amplifier setting during each experiment and correspond to electron charge transfer across the interface. Experiments for negative-polarity applied voltage, i.e., hole transfer across the interface, resulted in a similar series of current transients.

The initial spike in the transient corresponds to photogenerated charge separation in Se. A total charge equal to $C_{Se}V_{Se}$ is separated in the Se film during the laser pulse width, where C_{Se} and V_{Se} are the capacitance and voltage drop across the Se film, respectively. This charge can also be expressed as the product of the total cell capacitance C and the applied voltage V, i.e., $CV=C_{Se}V_{Se}$. The total time constant of the circuit, τ, was typically equal to 2 µsec which is reflected in the initial transient decay. The characteristic feature exhibited by the transients shown in Fig. 9 is the well-defined shoulder which would normally be associated with the transit time of charge carriers through the polymer. The "transit time" associated with the shoulder, however, exhibits a time dependence which is contrary to expected behavior. Instead of *decreasing*, it *increases*, as the applied electric field is increased! This behavior has been observed in fourteen different sandwich cells with a structurally similar configuration of Se and the polymer. The explanation of the observed lengthening of the "transit time" in the time-of-flight framework requires field-dependent mobility which would rapidly decrease as the field is increased.

To provide an explanation for these unusual results, we propose a phenomenological model based on the assumption that there exists a field-independent rate-limited charge-transfer mechanism at the Se-polymer interface. As a consequence the transient has a constant current component i_{pol} (Fig. 3b). This current flows in the polymer for a duration τ_d, which is the time necessary to deplete the charges accumulated in Se at the interface. Consequently the "depletion time" τ_d is proportional to the number of charges accumulated at the interface and therefore to the applied voltage. The electric field dependence of the depletion time τ_d for the transient currents shown above is presented in Fig. 10. For both electrons

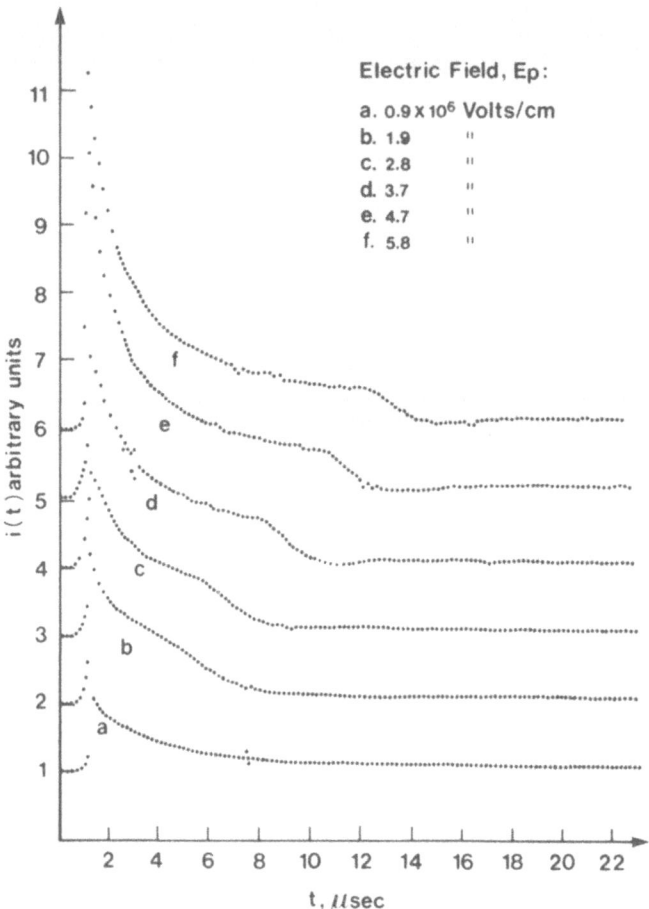

Fig. 9: A superposition of six photostimulated current transients
for applied electric fields across the polymer film between
0.9×10^6 V/cm and 5.8×10^6 V/cm. A positive voltage is
applied in this case which corresponds to electron charge
transfer across the Se-polymer interface.

and holes the depletion time varies linearly with the applied voltage and suggests that a rate-limited transfer of charge is taking
place. Another unique feature of the current transients is that
the current amplitude associated with charge transport in the polymer
for different applied voltages, is essentially constant. This
characteristic suggests that the charge-transfer rate at the interface is essentially independent of the applied electric field.

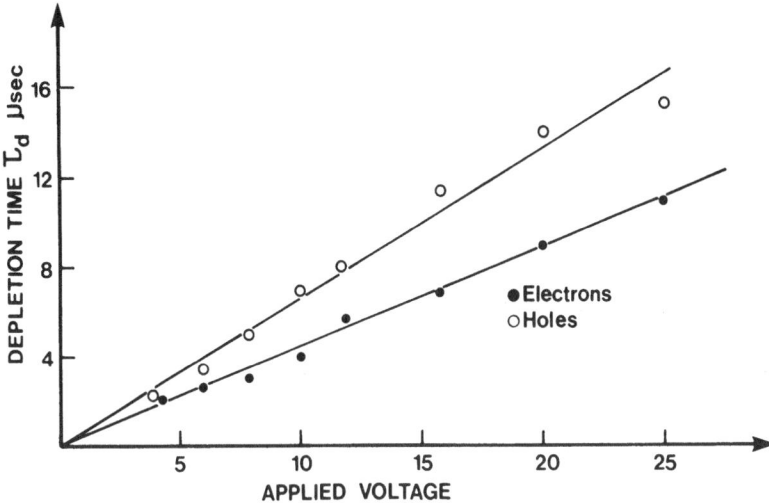

Fig. 10: A plot of the dependence of the depletion time τ_d as a
function of the applied voltage to the sandwich film.

If the phenomenological model is now extended to the portion
of the current transient beyond the depletion time, the current decay
at the end is associated with the "transit time", τ_t, of the charged
carriers through the polymer. At the time that the last charges are
transferred across the Se-polymer interface, the polymer film has
within it a uniform distribution of charges which are moving with
their characteristic mobility through the polymer under the influence
of the applied electric field. This induces a current in the external circuit which decays linearly, from the time defined at τ_d,
to zero as the charge carriers are transported across the polymer
film. In Fig. 11, the tails of four current transients for different
applied electric fields are superimposed at a common point corresponding to the depletion time of each transient. The variation in
slope as a function of the applied electric field is clearly seen.
There is a monotonic steepening in the current decay with increasing
applied electric field across the polymer indicating a corresponding
decrease in the transit time. As a result of the relatively long
time constant of the circuit it is difficult to determine the mobilities of the holes and electrons precisely from these decaying
currents. By approximately deconvoluting the effect of circuit

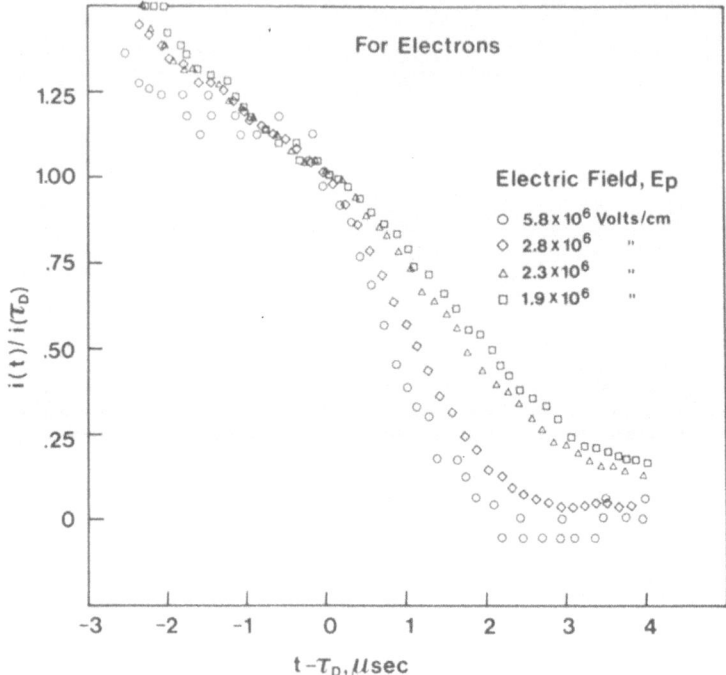

Fig. 11: A superposition of the tails of four current transients
          using the point corresponding to their depletion time
          as a common reference; for a variation of the applied
          electric field across the polymer from 1.9 x 10⁶ V/cm to
          5.8 x 10⁶ V/cm.

time constant, hole mobility has been estimated to be about $10^{-7} cm^2/V$
sec with electron mobility being half that of the holes. A similar
value for the mobility of holes in polystyrene has previously been
reported by Martin and Hirsch[14,15] and Kyokane et al[16]. This suggests that the copolymer of polystyrene and hexylmethacrylate has
charge-transport properties which are similar to the polystyrene
for holes and that the role of hexylmethacrylate is to support
electron transport.

    If indeed a charge-transfer rate-limited mechanism exists at
the interface, the depletion time measured in the current transient
should be dependent on the number of charges which are generated
and not the thickness of the polymer layer through which they are
transported. Using identical experimental conditions, the results of
experiments on two structurally similar sandwich cells were compared
as shown in Table 1. The difference between the two cells was in the

Table 1: Comparison between two structurally similar sandwich
          cells with copolymer thicknesses of 700 Å and 150 Å
          respectively.

	Sample A	Sample B
polymer thickness	700 Å	150 Å
Se thickness	900 Å	700 Å
C_{Se} (fds)	1.47×10^{-8}	1.89×10^{-8}
V_{Se} (volts)	11.6	9.0
Q_{Se} (coul.)	1.7×10^{-7}	1.7×10^{-7}
depletion time τ_d, holes	8×10^{-6} sec	9×10^{-6} sec
depletion time τ_d, electrons	5×10^{-6} sec	6×10^{-6} sec
E_{pol} (volt/cm)	3.2×10^{6}	3.4×10^{6}
applied voltage (volts)	34	14

thickness of their polymer layers, 700 Å and 150 Å, respectively.
The applied voltage to each sample cell was chosen such that the
product $C_{Se}V_{Se}$ was the same for both films and consequently the
same charge was generated at the Se–polymer interface. When a
comparison of depletion times was made between the two samples they
agreed within 15%, even though the thicknesses of their polymer
layers differed by a factor of 5. This gives strong support to a
rate-limited charge-transfer mechanism across the Se–polymer interface. Although the general conclusions about the rate-limited
charge-transfer mechanism at the Se–polymer interface explain the
main features of our experimental results, the microscopic model is
uncertain at this time.

One aspect of the transfer of charge across the Se-polymer interface which is most pertinent to the imaging process in the particle
migration microfilm is the charge transfer efficiency for both holes
and electrons. This was determined by deconvoluting the current
transients into two parts; one associated with charge separation
within the continuous Se film, labelled i_{Se} in Fig. 2b), and the
other corresponding to charge transport through the copolymer film,
labelled i_{pol}. By integrating the deconvoluted current transients
using a correction for the dielectric thicknesses of both the Se and
copolymer films, the charge transfer efficiency across the interface
could be evaluated. In Fig. 12, the ratio of the charge which was
transported across the polymer to the charge which was photogenerated
in Se, is plotted against a range of applied electric fields in the

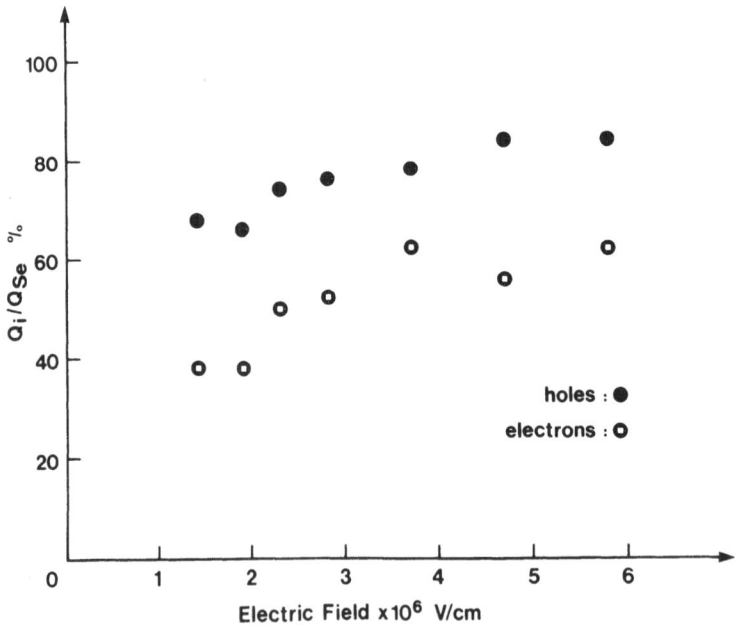

Fig. 12:   Charge transfer efficiency across the Se-copolymer
           interface for holes and electrons as a function of
           applied electric field across the polymer.

polymer. The results show that, over the entire electric field range
studied, hole charge transfer across the interface is 20% more efficient than electrons.

In summary, a new charge-transfer mechanism has been discovered
at the interface of continuous sandwich film structures composed of
amorphous Se and polystyrene hexylmethacrylate copolymer. It was
concluded that the new mechanism is associated with interfacial states
or sites which transfer charge across the Se—polymer interface according to an essentially field-independent rate-limited process. As a
consequence, photogenerated current transients exhibit a constant-
current contribution associated with the depletion of photogenerated
charge which is accumulated in Se at the interface. This constant
current begins to decay when charge depletion in Se is complete, i.e.,
after the "depletion time" τ_d.

DISCUSSION

Photostimulated current transient measurements were applied to
electroded particle migration microfilm, as well as continuous thin
film structures of Se and styrene hexylmethacrylate, in order to
elucidate the charge transport phenomena during the imaging process
in the film. It was determined that illumination of a film, across
which an electric field has been applied, results in separation of
photogenerated charge in the selenium spheres which are located just
below the surface of the film. Charge injection from Se into the
polymer, which proceeds by a rate-limited field-independent process,
takes place immediately following charge separation in the Se sphere
for electric fields greater than 10^4 Volts/cm. From the experimental
results for continuous thin film structures of Se and the copolymer,
it was determined that the charge transfer efficiency across the Se-
copolymer interface is 20% higher for holes than for electrons in the
high field range. Therefore, it may be concluded that the Se sphere
becomes negatively charged during the imaging process.

ACKNOWLEDGEMENTS

The authors wish to thank Zoran Popovic and George Hartmann for
useful discussions, as well as Khamo Oraha for his technical assistance.

REFERENCES

1. U.S. Patent 3,502,681, Photoelectrosolography, W.L. Goffe (to
    Xerox Corporation).
2. W.L. Goffe, Photogr. Sci. Eng., 15, 304 (1971).
3. A.L. Pundsack, Photogr. Sci. Eng., 18, 642 (1974).
4. A.L. Pundsack, Y.C. Cheng, G.C. Hartmann and L.M. Marks, Appl.
    Optics, 17, 2650 (1978).
5. K.M. Hong, J. Opt. Soc. Am., 70, 821 (1980).

6. S. Tutihasi, Photogr. Sci. Eng., 18, 394 (1974).
7. D.A. Buckley and P.P. Augostini, Polymer Reprints, 18, 528 (1977).
8. Hy-Sil Manufacturing Inc., Revere, Mass. 02151.
9. C.C. Yang, J.Y. Josefowicz and L. Alexandru, Thin Solid Films,
    in press.
10. J.L. Hartke, Phys. Rev., 125, 1177 (1962).
11. W.E. Spear, Proc. Phys. Soc., B 70, 669 (1957); B 76, 826 (1960).
12. J.L. Hartke and P.J. Regensburger, Phys. Rev., 139, A 970 (1965).
13. J.Y. Josefowicz, C.C. Yang and Z. Popovic, Phys. Rev. Letters,
    43, 886 (1979).
14. E.H. Martin and J. Hirsch, Solid State Commun., 7, 783 (1969).
15. E.H. Martin and J. Hirsch, J. Appl. Phys., 43, 101 (1972).
16. J. Kyokane, S. Harada, K. Yoshino and Y. Inuishi, Jap. J. of
    Appl. Phys., 18, 1479 (1979).

HIGHLY CONDUCTING POLY(p-PHENYLENE) VIA SOLID-STATE

POLYMERIZATION OF OLIGOMERS

L.W. Shacklette, H. Eckhardt, R.R. Chance, G.G. Miller,
D.M. Ivory, and R.H. Baughman

Corporate Research Center
Allied Chemical Corporation
Morristown, New Jersey 07960

We have discovered a novel method of preparation of highly conductive polymers: the simultaneous solid state polymerization and doping of poly(p-phenylene) oligomers by the action of strong Lewis acids such as AsF_5. We have previously shown that higher molecular weight poly(p-phenylene), PPP, can be doped with either strong electron acceptors (AsF_5, IF_5, HSO_3F, SO_3, $SbCl_5$) or donors (Na, K, Li) to form highly conducting complexes.[2,3] In the present case, we have observed that biphenyl, p-terphenyl, p-quaterphenyl, p-quinquephenyl and p-sexiphenyl polymerize and dope in the presence of AsF_5 to form acceptor-doped metallic poly(p-phenylene). We have also observed that the donor dopant, potassium reacts with p-oligophenylenes to form highly conductive complexes, although in this case there is no evidence for increased chain length.

This discovery of a gas-solid polymerization reaction is particularly significant because it allows the use of the highly processible oligomers of polyphenylene as precursors to the insoluble, infusible poly(p-phenylene). In this way, thin conductive films of metallic poly(p-phenylene) can be prepared directly from films of the oligomers which have in turn been deposited from the vapor phase or from solution. Alternatively, chain-oriented poly(p-phenylene) can be produced by reacting single crystals or epitaxially grown films of the oligomers with AsF_5.

Powders, films, and single crystals of the polyphenylene
oligomers ($2 \leq n \leq 6$) have been reacted with 400-Torr AsF$_5$ at room
temperature for periods of up to 24 hours. In all cases a highly
conductive material results, although the highest conductivities
have been obtained with p-terphenyl as the starting material.
In this case conductivities as high as 50 S/cm have been measured.
Elemental analysis of the product obtained from the terphenyl-
AsF$_5$ reaction typically gives: $C_6H_{4.42}$ (AsF$_4$)$_{.2}$, which indicates
a C/H ratio (1.36) between the theoretical ratio of high
molecular weight polyphenylene (1.5) and terphenyl (1.29). Also,
elemental analysis of high molecular weight poly(p-phenylene)
doped with AsF$_5$ results in a F/As ratio of 5 instead of the 4
observed in the present case. Together these facts suggest that
AsF$_5$ couples terphenyl molecules via a sequence of steps involving
fluorination of the terminal phenyl rings and elimination of HF.
The loss of HF would explain the low F/As ratio found in the
polymerized material.

Evidence for the polymerization of terphenyl is given in
Fig. 1 which compares the IR spectrum of the annealed product
of the terphenyl-AsF$_5$ reaction with those of p-terphenyl (n=3)
and poly(p-phenylene) (n$\sim$16), the latter being prepared by the
method of Kovacic et al.;[4,5] annealing the terphenyl-AsF$_5$ product
sublimes away unreacted terphenyl and the compensated dopant.
All samples shown in the figure are insulating; the variation
of background absorption is due to scattering. The relative
strength of the absorption bands near 800 cm^{-1} and 765 cm^{-1},
due respectively to the C-H out-of-plane vibrations of para-
and mono-substituted phenyl rings, can be used to estimate
chain length.[5,6] By this method we estimate the average chain
length ($\bar{n}$) of reacted and annealed terphenyl to be 9 monomer
units (i.e., a mixture of n=6,9,12, etc. whose average is 9).
Reactions using biphenyl as the starting material yield
similar results for $\bar{n}$. Reactions with sexiphenyl have produced
average chain lengths as long as 16 to 18. We have also
previously reported evidence for further polymerization induced
by AsF$_5$ doping of samples containing predominantly long-chain
poly(p-phenylene).[3] In this case, coupling between polymer chain
ends is possible only for those chains whose ends happen to lie
close together.[1] Since the rod ends of the oligomers are held
in relative proximity by the requirements of a three dimensional
crystal structure, numerous couplings along a given chain are
possible. Although we have presently not achieved chain lengths
longer than 16, the process of polymerization in a crystalline
matrix can potentially lead to higher molecular weight polymer
than can be achieved by previously used synthetic methods.

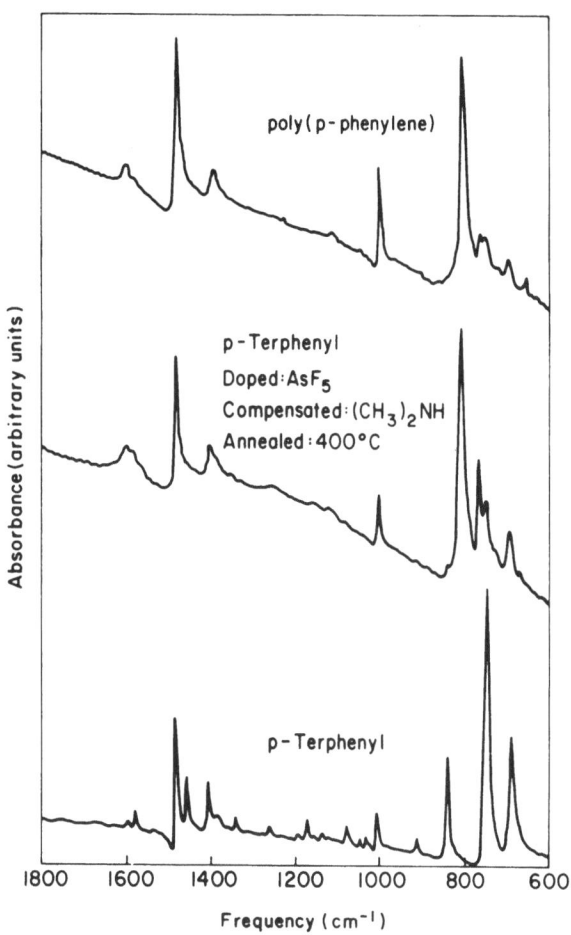

Fig. 1.   A comparison of IR spectra for p-terphenyl,
poly(p-phenylene) and reacted p-terphenyl after the
latter was treated with AsF$_5$, (CH$_3$)$_2$NH and annealed
at 400°C (a temperature where compensated dopant and
unreacted terphenyl is sublimed). All such treated
samples are insulating. The variation of background
absorption is due to scattering.

The polymerization of terphenyl and quaterphenyl has also been investigated by means of X-ray diffraction. The results of this study are presented in Table 1 which compares the d spacings obtained for terphenyl, terphenyl and quaterphenyl reacted with AsF_5, and poly(p-phenylene). After reaction with AsF_5, the washing and annealing process removes all traces of unreacted terphenyl or quaterphenyl leaving a product which represents about 2/3 of the starting weight of oligomer and has an X-ray diffraction pattern nearly identical to that of poly(p-phenylene). One may note from Table 1 that three of the lattice spacings exhibited by terphenyl correspond to the set of spacings found for poly(p-phenylene). This correspondence results from the fact that the crystalline polymer exhibits the same lateral packing between chains as the phenylene oligomers exhibit between molecular rods.[3] In addition, the transverse distances between chains are approximately the same as the corresponding spacings in the single crystalline oligomers. The other four spacings given for terphenyl depend on the length of the terphenyl molecule for which there is no crystallographic analog in the polymer.

TABLE 1:   INTERPLANAR DISTANCES (Å) FOR PHENYLENE COMPOSITIONS

Terphenyl	Terphenyl, AsF_5, $(CH_3)_2NH$ Annealed 400°C	Quaterphenyl, AsF_5, $(CH_3)_2NH$, Annealed 400°C	PPP Undoped n ~16	PPP AsF_5 $(CH_3)_2NH$ Annealed 400°C
13.4 (s)	Spacings Related to the Length of the Terphenyl Molecule			
6.76 (m)				
4.60 (s)	4.53 (s)	4.51 (s)	4.48 (s)	4.48 (s)
4.52 (s)	Spacings Related to the Length of the Terphenyl Molecule			
4.34 (m)				
3.86 (m)	3.93 (m)	3.90 (m)	3.89 (m)	3.90 (m)
3.17 (m)	3.19 (m)	3.20 (m)	3.17 (m)	3.19 (m)

Evidence for the polymerization of terphenyl can also be found in the UV-visible transmission spectra of thin terphenyl films grown on a quartz substrate. These films were mounted in a stainless-steel cell equipped with quartz windows. The evolution of the spectra upon the introduction of 20-Torr AsF_5 is presented in Fig. 2. The spectrum of pure undoped p-terphenyl exhibits a strong absorption peak at approximately 4.5 eV. This absorption, which is associated with the π to π^* transition has been showed by previous workers[6,7,8] to shift toward the

visible as chain length increases. The results which are of
particular importance here are for p-terphenyl (4.5 eV),
p-sexiphenyl (3.9 eV), and poly(p-phenylene) (3.14 to 3.35 eV
measured by previous workers[6], 3.43 in the present work, from
diffuse reflection for polymer prepared by the Kovacic method).
Fig. 2 gives clear evidence for a shift in the peak absorption
from 4.5 eV to 3.6 eV during exposure to AsF_5. This shift toward
longer wavelength is a permanent feature which is not reversed
by compensation with $(CH_3)_2NH$. The evidence of Fig. 2 suggests
a polymerization to chain lengths significantly longer than 6.

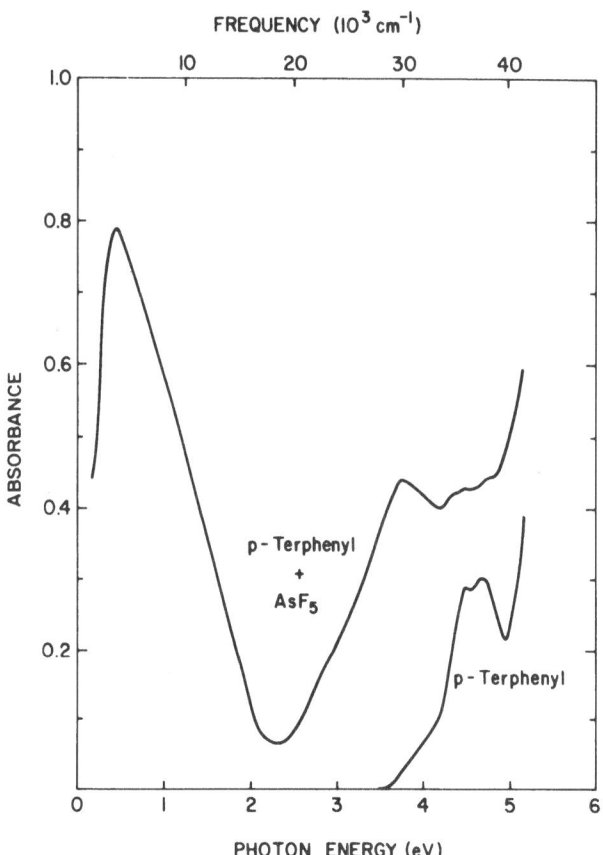

Fig. 2.   The change in the transmission spectrum of a film
          (10^{-3} cm) of p-terphenyl upon exposure to AsF_5.   The
          film is microcrystalline; transmission is perpendicular
          to the ab face.   The UV-visible, near IR, and IR
          frequency ranges were taken on separate samples.   The
          data were normalized in their respective regions of
          overlap, starting with the UV-visible range.

Since the position of the UV-reflection maximum is indicative
of chain lengths for the p-phenylenes, it is important to have
careful measurements of the position of this maximum for polymer
prepared by various methods. Previously reported literature
values range from 3.14 eV (395 nm)[6] to 3.65 eV (340 nm).[9] A value
of 344 nm (3.61 eV) is calculated for infinite - chain poly
(p-phenylene) from oligomer data and Kuhn's root law.[10]

The commercially available instruments for measuring diffuse
reflectance give very broad featureless maxima which make precise
determination of λ_{max} extremely difficult. We have used a more
flexible system which employs separate excitation and collection
monochromators and crossed polarizers to help reduce the amount
of specularly reflected light collected by the detector. A detailed description of the apparatus is given elsewhere.[11] We
also found that it was helpful to cool the powdered polymer to
LN_2 temperature for grinding in order to obtain sufficiently
fine particles. The powdered polymer was then mixed with barium
sulfate and the reflectance of this mixture was compared with a
standard of pure barium sulfate. This method gave relatively
sharp peaks (width at half maximum ∿100 nm) with clearly defined
maxima. We obtained 362 nm (3.43 eV) for the dark brown poly
(p-phenylene) prepared by the Kovacic method (chain length
16-18 phenyl units inferred from IR). Light yellow polymer
prepared by the method of Yamamoto et al.[12] exhibited a maximum
at 345 nm (3.59 eV) (chain length 14 to 16 units from IR).
This polymer also fluoresced in the visible making the dual mono-
chromator arrangement indispensable. The IR spectra suggest
that the difference in λ_{max} between the two polyphenylenes is
not due to differences in chain length, since they both appear
to have a comparable density of mono vs. para substituted phenyls.
ESR measurements on the two polymers indicate a significantly
higher concentration of radicals in the brown polymer. These
radicals, which will have a quinoid structure, could contribute
to a higher λ_{max} and to the absorption tail in the visible
which is responsible for the brown color.

Again returning to Fig. 2 the rising absorption toward the
red end of the spectrum is likely due to free carriers in the
doped (metallic) polymer. This feature is rapidly reversed by
exposure to $(CH_3)_2NH$. Absorbance in the doped polymer continues
to increase with decreasing frequency until about 3600 cm^{-1},
beyond which the absorbance levels off and declines. In a simple
Drude model for free carriers, one expects that conductivity
(to which absorption coefficient is related) will begin leveling
off to its approximate dc value at a frequency where $\omega\tau\sim1$
(where ω is the frequency of the exciting field and τ is the
relaxation time). A frequency independent conductivity leads
through Maxwell's equations to an absorption coefficient which
declines as $\sqrt{\omega}$ toward low frequency. The decrease in ab-

sorbance at very low frequency could also be the result of an
inhomogeneous (or granular) distribution of the metallic phase.
In this case there are two effects which can lead to the observed
decrease in absorption. In the first case, the absorption
coefficient is reduced when interparticle resistance sharply
lowers conductivity as wavelength becomes comparable to particle
size. In the second case, the film becomes more transparent
when the intensity of scattered light is reduced as wavelength
becomes larger than particle size.

        As was mentioned in the introduction the doping of
crystalline oligomers can lead to conductive chain-aligned
poly(p-phenylene). An illustration of this possibility is provided by Fig. 3 which shows the doping and polymerization
of a slab of p-terphenyl cut from a single crystal boule where
conduction is approximately parallel and perpendicular to the
original c-axis of terphenyl. The anisotropy in the electrical
conductivity is evidence for chain alignment, although the
degree of anisotropy is not as great as that observed in polyacetylene.[13]

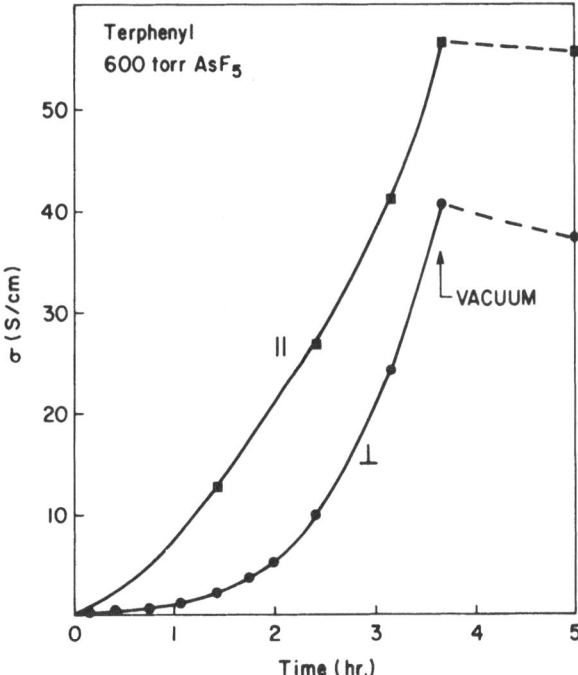

Fig. 3.   The growth of anisotropic conductivity in a single crystal
           slab of p-terphenyl (approx. parallel and perpendicular
           to the original c-axis of terphenyl).

We have also reacted terphenyl and sexiphenyl with potassium metal to form conductive complexes. Since previous experiments on poly(p-phenylene)[3] have indicated a maximum donor dopant level of approximately one potassium atom for every two monomer units, the oligomers in this experiment were exposed to potassium metal in the same ratio (0.5K/monomer). A small (catalytic) amount of naphthalene was added to the phenylene-potassium mixture and tetrahydrofuran (THF) was distilled into the reaction vessel under vacuum conditions. The mixture was then stirred for 12 hours until the reaction was complete. The chamber was placed under dynamic vacuum for another several hours in order to remove the THF. Pellets were then pressed from the remaining black residue and conductivities were measured with a 4-in-line probe in an argon filled dry box. Freshly pressed pellets of the potassium-doped sexiphenyl had a brilliant gold appearance which tarnished rapidly even inside the dry box. Similar pellets of K-doped terphenyl had a greenish gold color and were also extremely unstable in O_2 and H_2O vapor. Infrared spectra taken on compensated samples indicate that no increase in chain length occurs upon doping with potassium. A comparison of chain length vs conductivity is presented in Table 2. It is satisfying to note that conductivity increases with chain length, but interpretation is complicated by the fact that chain length may affect the conductivity in a number of ways including interchain mobility, intrachain mobility, degree of charge transfer, degree of charge localization, and stability.

TABLE 2:   CONDUCTIVITIES OF POTASSIUM-COMPLEXED p-PHENYLENES

Chain Length n	Conductivity σ (S/cm)	Appearance
3	4×10^{-5}	yellow-green
4	2×10^{-5}	green-gold
6	0.5	Gold
16	7	Gold

The processibility of the phenylene oligomers potentially leads to several interesting applications, among which is the possibility of fabricating Schottky photojunctions using thin films of metallic poly(p-phenylene). We have already constructed Schottky barrier diodes using n-type titanium dioxide and p-type gallium phosphide against metallic poly(p-phenylene) which indicate a very high metallic work function for degenerately doped PPP (5.4 eV). A high work function potentially allows one to construct photojunctions displaying large open circuit voltages.

References

1. L. W. Shacklette, H. Eckhardt, R. R. Chance, G. G. Miller, D. M. Ivory, and R.H. Baughman, J. Chem. Phys., in press; R. R. Chance, L. W. Shacklette, G. G. Miller, D. M. Ivory, and R. H. Baughman, Bull. Am. Phys. Soc. 25, 399 (1980).

2. D. M. Ivory, G. G. Miller, J. M. Sowa, L. W. Shacklette, R. R. Chance, and R. H. Baughman, J. Chem. Phys. 71, 1506 (1979).

3. L. W. Shacklette, R. R. Chance, D.M. Ivory, G. G. Miller, and R. H. Baughman, Synth. Met. 1, 307 (1980).

4. P. Kovacic and A. Kyriakis, J. Am. Chem. Soc. 85, 454 (1963).

5. P. Kovacic and J. Oziomek, J. Org. Chem. 29, 100 (1964).

6. P. Kovacic and L. Hsu, J. Polym. Sci. Part A1, 4, 5 (1966).

7. A. E. Gillam and D. H. Hey, J. Chem. Soc. (London), 1939, 1170.

8. I. B. Berlman, "Fluorescence Spectra of Aromatic Molecules" (Academic Press, 1971).

9. H. F. VanKerckhoven, Y. K. Gillam, and J. K. Stille, Macromolec. 5, 541 (1972).

10. W. Kuhn, Helv. Chim. Acta, 31, 1780 (1958).

11. R. R. Chance and J. M. Sowa, J. Am. Chem. Soc., 99, 6703 (1977).

12. T. Yamamoto, Y. Hagashi, A. Yamamoto, Bull. Chem. Soc. of Japan, 51, 2091 (1978).

13. Y. W. Park, M. A. Druy, C.K. Chiang, A. G. MacDiarmid, A. J. Heeger, H. Shirakawa, and S. Ikeda, J. Polym. Sci., Polym. Lett. Ed. 17, 195 (1979).

CONDUCTING COMPLEXES OF A PROCESSIBLE POLYMER:

POLY(p-PHENYLENE SULFIDE)

R.R. Chance, L.W. Shacklette, H. Eckhardt, J.M. Sowa,
R.L. Elsenbaumer, D.M. Ivory, G.G. Miller, and
R.H. Baughman

Corporate Research Center
Allied Chemical Corporation
Morristown, New Jersey 07960

INTRODUCTION

The discovery that polyacetylene, $(CH)_x$, can be doped with electron donors or electron acceptors to yield highly conducting derivatives[1] has resulted in a great deal of interest in doped polymer systems and their potential for commercial application as replacements for semiconductors or metals. This interest was heightened by the discovery of two additional conducting polymer systems based on polypyrole[2] and poly(p-phenylene) [PPP].[3] However, none of these polymers is either melt or solution processible, an important consideration for the majority of potential commercial applications.

Two polymers have been recently reported which are both melt and solution processible and which can be doped to form conducting complexes.[4-6] Poly(m-phenylene) can be doped with strong electron acceptors such as AsF_5 to yield a moderately conducting complex ($\sigma \sim 10^{-3}$ S/cm).[4] The second system, which has been simultaneously reported by Chance et al.[5] and Rabolt et al.,[6] is based on poly(p-phenylene sulfide), PPS. This high molecular weight polymer is melt and solution processible and is available commercially from Phillips Petroleum Company. This paper will concentrate on the electrical and optical properties of doped PPS. Possible chemical modification of PPS on doping will also be discussed.

ELECTRICAL AND OPTICAL PROPERTIES

On exposure to AsF_5, the conductivity, σ, of PPS increases by about 16 orders of magnitude to about 1 S/cm. Typical results for a pressed pellet of Ryton type V-1 PPS powder obtained

from Phillips are shown in Figure 1. The initial increase in
σ is very rapid as the surface is doped and is followed by a more
gradual increase, which is probably due to the kinetics of the
doping process being limited by diffusion into the bulk of the
sample. After prolonged exposure to AsF$_5$, σ ~1 S/cm has been
achieved for 0.025 mm thick films, 0.025 mm diameter fibers, and
1 mm thick pressed pellets of PPS. This σ value is about 1000
times lower than that obtained for AsF$_5$-doped (CH)$_x$[1] or PPP[3] and
1000 times greater than that obtained, thus far, for AsF$_5$-doped
poly(m-phenylene). Donor doping of PPS with a potassium
naphthalide solution, which has produced high conductivities in
previous systems[1,3], yields a blue-black complex with a low
conductivity (σ ~10^{-6} S/cm).

The conductivity of AsF$_5$-doped PPS is more strongly temperature dependent than that for AsF$_5$-doped (CH)$_x$ or PPP, as might
be expected given the reduced σ values in the PPS system. Acti-

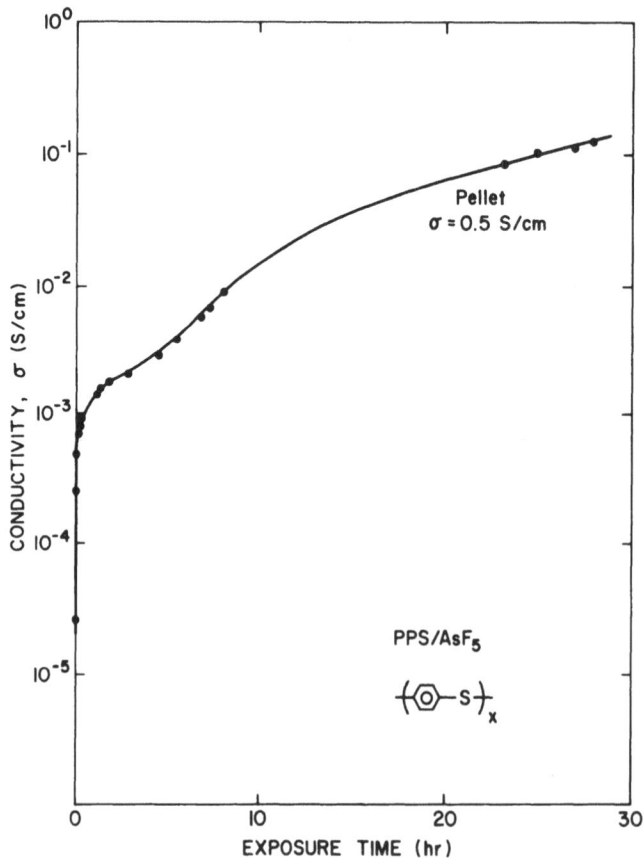

Fig. 1.   Electrical conductivity versus time of exposure to
           AsF$_5$ at 300 torr for PPS pressed pellet.

vation energies derived from plots of log σ vs. T^{-1} near room
temperature are about 0.05 eV for our most heavily doped samples
(approximately 1:1 ratio of AsF_5 to C_6H_4S) and about 0.5eV for
the most lightly doped samples. The temperature dependence can
be adequately represented by plots of log σ vs $T^{-1/2}$ as shown
in Figure 2. This behavior is consistent with trap-modulated
metallic conductivity in one dimension,[7] but has also been observed for highly conducting particles imbedded in an insulating
matrix.[8]

As in other systems,[1,3] the thermoelectric power of AsF_5-
doped PPS is positive (+ 40 $\mu V/°K$ at room temperature) indicating
mobile hole-like carriers, i.e., a p-type conductor. The conductivity can be compensated by addition of electron donating
agents such as dimethylamine or NH_3. Junction measurements in a
Schottky barrier configuration are also consistent with the
p-type character of AsF_5-doped PPS.

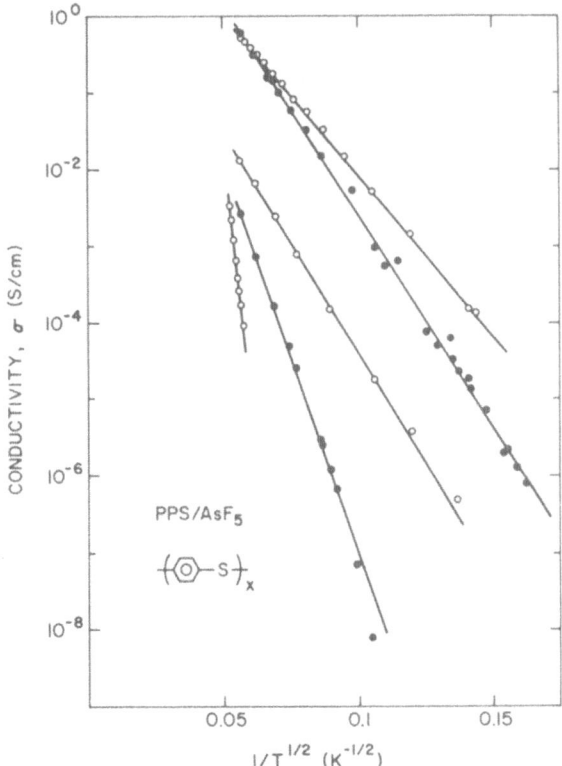

Fig. 2.   Electrical conductivity versus $T^{-1/2}$ for AsF_5-doped
          PPS (Pressed pellets) at various doping levels.
          Doping level increases from left to right in the
          figure.

The variations in optical absorption on doping with AsF_5 over a broad spectral range (IR to UV) are illustrated in Figure 3. The conductivity has been monitored simultaneously and the approximate values appropriate for the spectra are indicated in the figure. The AsF_5 exposure conditions are restricted so that relatively low σ values (<10^{-3} S/cm) are achieved, since heavy doping results in opaque films for the film thicknesses presently available (≥0.025 mm).

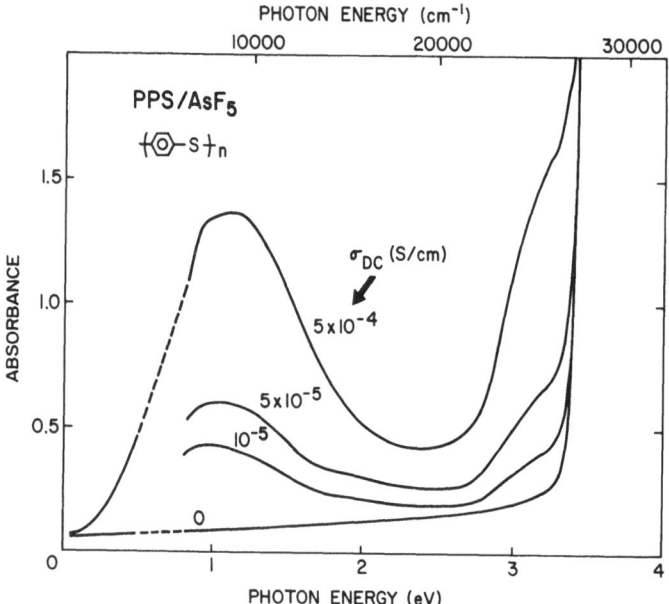

Fig. 3.   Variation in optical absorption of a 0.025 mm thick
          film of PPS on doping with AsF_5 (30 torr).   IR
          absorption for the polymer vibrations, which extend
          to about 0.7 absorbance units, are omitted because
          of the compressed energy scale.

The dominant feature of the spectra is a broad peak centered at about 1.1 eV with a width of ∿1 eV. The large decreased in absorption at low energies is consistent with the lower d.c. conductivities in AsF_5-doped PPS compared to other systems which show much stronger IR absorption.[1,3] There are several possible explanations for the 1.1 eV peak (and the low σ values in PPS). First, AsF_5-doped PPS may be a semiconductor, in which case the 1.1 eV peak would map out a convolution of the PPS valence band with the distribution of unoccupied AsF_5 acceptor levels in the gap. Second, the AsF_5-doped PPS may be a inhomogeneous system consisting of metallic particles imbedded in an insulating (or semiconducting) matrix, in which case the high energy side of the 1.1 eV peak could be attributed to free-carrier absorption and the decrease in absorption at low energies attributed to a particle size effect. As mentioned previously, the temperature dependence of σ is consistent with the latter. Third, the same general behavior of absorption coefficient and conductivity can also be expected under certain conditions for a disordered one-dimensional metal.[7] Fourth, even in a simple Drude model for free carriers, one expects that the absorption coefficient will decrease with decreasing frequency once $\omega < 1/\tau$ where τ is the relaxation time. To attempt to fit the curve of Figure 3 one would require a very short relaxation time compared to that for a good metal. The Drude model, however, predicts a comparatively rapid fall off at low frequency and provides a very poor fit to the nearly symmetric 1.1 eV peak of Figure 3.

A shoulder on the main absorption band of PPS (located at 3.6 eV) develops at about 3.3 eV during the doping. It is tempting to attribute this absorption to transitions from occupied acceptor states in the gap to the conduction band (in the semiconductor model discussed above). However, we believe this peak is due to a chemical modification of the polymer, since the 3.3 eV peak, unlike the 1.1 eV peak, is not completely eliminated on compensation.

Preliminary experiments have been conducted on uniaxially oriented PPS films. The degree of orientation obtained from X-ray measurements on these films is very high. Reflection spectra in the 25,000 to 40,000 cm^{-1} range are consistent with a highly oriented film in that polarization parallel to the stretch direction is highly structured while perpendicular polarization yields a completely structureless spectrum. This result also demonstrates that the undoped material is highly one-dimensional in spite of the nonplanar backbone structure[9] illustrated in Figure 4. This suggests that the nearly per-pendicular arrangement of phenyl rings is near optimal for electronic delocalization in the phenylene sulfide system. The maximum conductivity anisotropy observed thus far for these films is about a factor of six at low doping levels. The anisotropy for optical absorption in the near infrared region is less than a factor of two. Thus, even though the degree of alignment for stretched PPS films is undoubtedly much greater than that of stretched polyacetylene films,[2] the apparent conductivity anisotropy is considerably less than the factor of fifteen observed for polyacetylene.[1] However, in the polyacetylene case the fibular nature of the material could lead to extrinsic

contributions to the conductivity anisotropy due to interparticle
resistance effects.   Also, it should be emphasized that our
anisotropy measurements on PPS have thus far been restricted
to low doping levels.

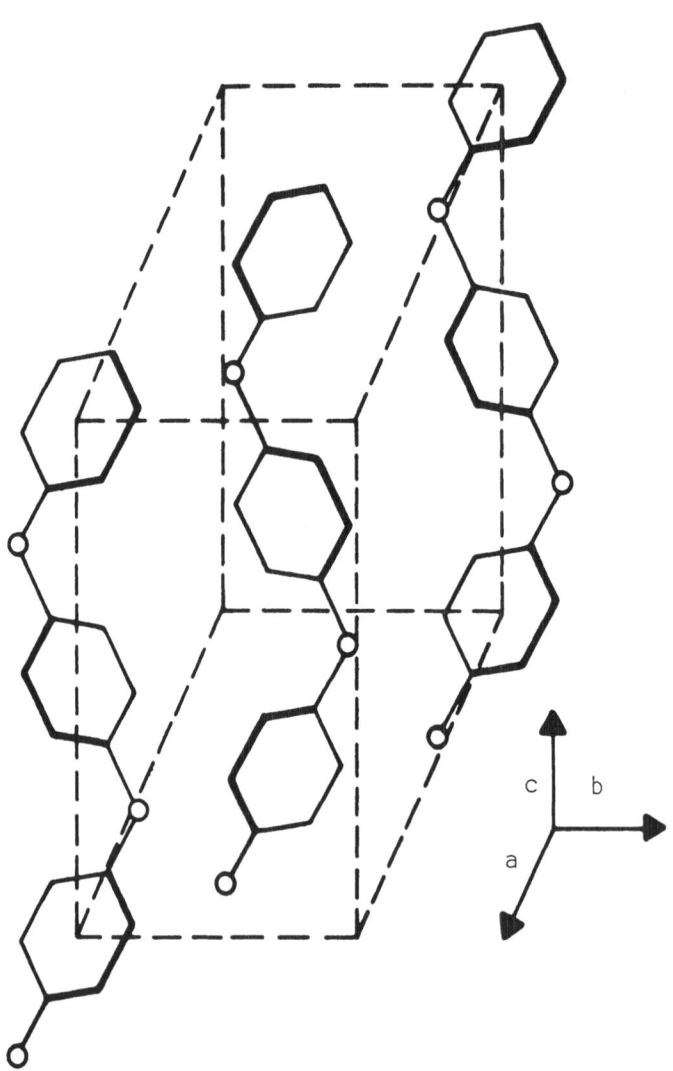

Fig. 4.   Structure of PPS from reference 9.

STABILITY

An important consideration for most commercial applications
is the stability of the doped material on exposure to normal
environmental conditions. We find that the conductivity of
AsF_5-doped PPS is less stable than either $(CH)_x/AsF_5$ or
PPP/AsF_5 on exposure to laboratory air. The major cause of this
instability is water vapor in the air. Exposure to dry O_2
(300 torr for 48 hours) has little effect on σ while the introduction of small amounts of water vapor (15 torr) yields a
precipitous σ decrease.

PPS/AsF_5 is quite stable under high vacuum conditions
($\sim 10^{-6}$ torr) at room temperature. However, at higher temperatures
($\sim 50°C$) a degradation of σ begins. For PPS/AsF_5 films, the
degradation rate $D = \sigma^{-1} d\sigma/dt$, varies roughly as $t^{-\frac{1}{2}}$ under
isothermal conditions. This result probably indicates a loss
of dopant with the kinetics limited by diffusion to the surface.
For small overall variations in σ, the temperature dependence
of D can be satisfactorily represented by an Arhenius expression
with an activation energy of ~ 0.7 eV. Typically we find
$D = 0.01$ hr^{-1} at 50°C.

CHEMICAL MODIFICATION

Elemental analysis of virgin PPS typically yields results
very close to the theoretical composition $C_6H_4S_1$. Thermal
analysis yields: $T_g = 85°C$, $T_c = 135°C$ and $T_m = 280°C$ from DSC and
only a 1% weight loss up to 400°C from TGA. Elemental analysis
of AsF_5-doped PPS (σ = 1-4 S/cm) yields results in the range
$C_6H_4S_{1.0}$ $(AsF_{4.3-6.6})_{0.75-0.81}$. Somewhat disturbing is the
variable F/As ratio. After exposure of the conductive samples
to either ammonia or dimethylamine, which rapidly compensates
the conductivity, followed by extensive water washing to remove
inorganics, the elemental analysis yields results in the range
$C_6H_{2.7-2.9}S_{1.0}F_{0.2-0.35}$. (Maximum N and As levels detected
were 0.5%). This result suggests some fluorination of the
phenyl rings and substantial crosslinking (to account for the
lowered H). DSC reveals no thermal transitions for this
material.

Clearly, the process of doping and compensation results
in considerable chemical change in the polymer. This conclusion has been confirmed by infrared studies shown in Figure 5.
At relatively low molar concentrations of AsF_5, $[AsF_5] < 0.5$,
(per mole of C_6H_4S), there is little indication of
chemical change in the infrared spectrum. At higher $[AsF_5]$,
the infrared indicates a substantial chemical modification.
In particular the modes associated with para substituted phenyl
rings are substantially decreased in intensity and new bands
appear which indicate multisubstituted phenyl rings and probably

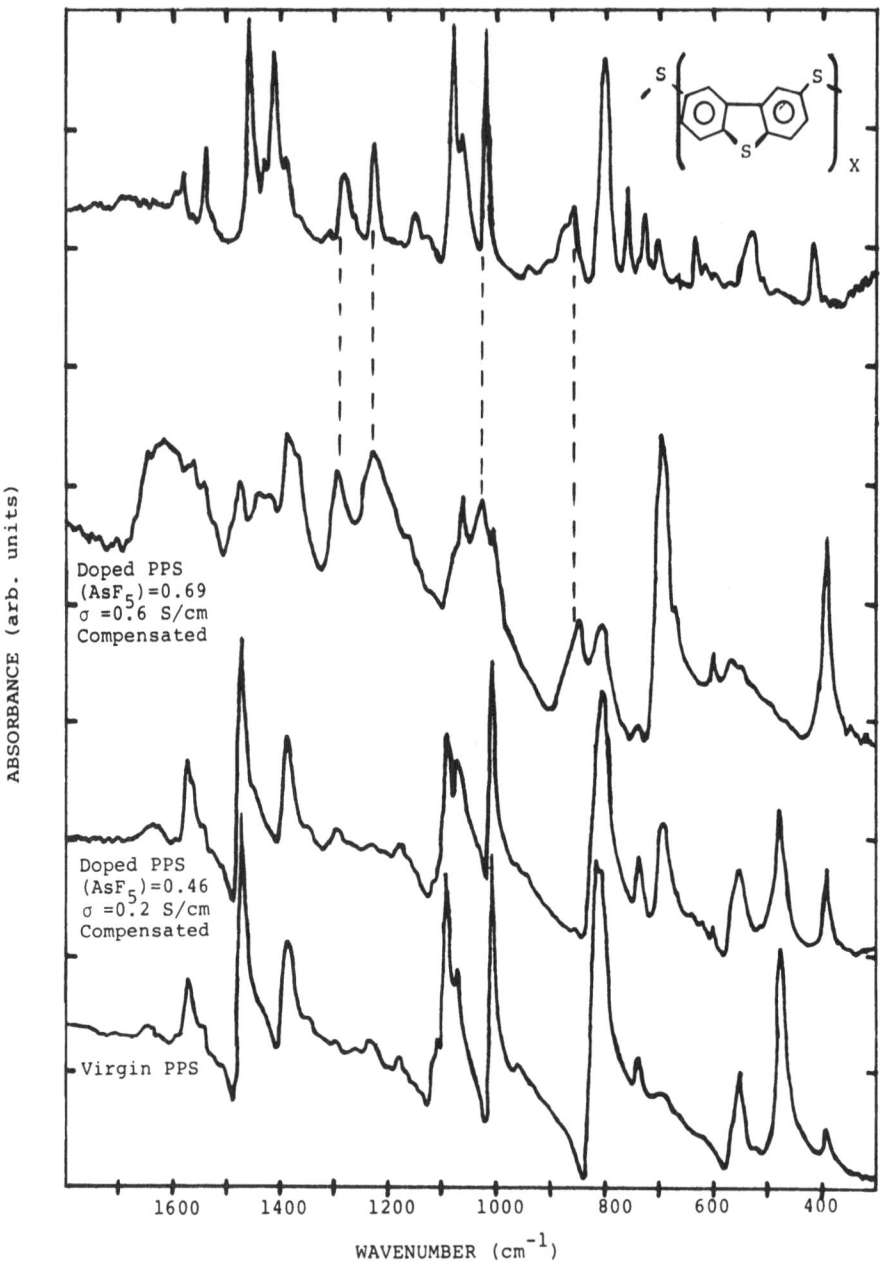

Fig. 5.   Infrared spectra for virgin PPS; doped, compensated,
and washed PPS at doping levels of 0.46 and 0.69;
and polydibenzothiophene sulfide (PDTS). The peaks
at 680 cm^{-1} and 395 cm^{-1} in the doped PPS spectra
are due to AsF$_6^-$. The dashed vertical lines
indicate the major peaks which are absent in
virgin PPS but present in PDTS and in heavy doped
and compensated PPS. The compensating agent for
these samples was dimethylamine.

some fluorination of the phenyl rings (broad band at 1290 cm^{-1}). According to the elemental analyses discussed above, there is not enough fluorine present to account for the lowered H content. Therefore, we believe the chemical modification involves mainly some type of crosslinking. Certainly, the loss of the thermal properties of the parent polymer after heavy doping and compensation would be consistent with substantial crosslinking.

There are two possibilities for crosslinking: intramolecular to yield, for example, poly(dibenzothiophene sulfide) $(C_{12}H_6S_2)_x$ (PDTS) and intermolecular to yield a randomly crosslinked polymer. An infrared spectrum for PDTS is also shown in Figure 5 and shows a rough correspondence with the new bands which appear on doping and compensation.

A remaining question is whether the crosslinking is a limiting factor with respect to achievable conductivities or whether chemical modification provides a new material (PDTS, for example) which becomes conductive with AsF$_5$ doping. In an attempt to address this question, we have carried out AsF$_5$ doping experiments at various temperatures in the hope that the kinetics of doping and the kinetics of the crosslinking process would display different temperature dependences. The results are summarizared in Figure 6 and Table I. The 50°C and

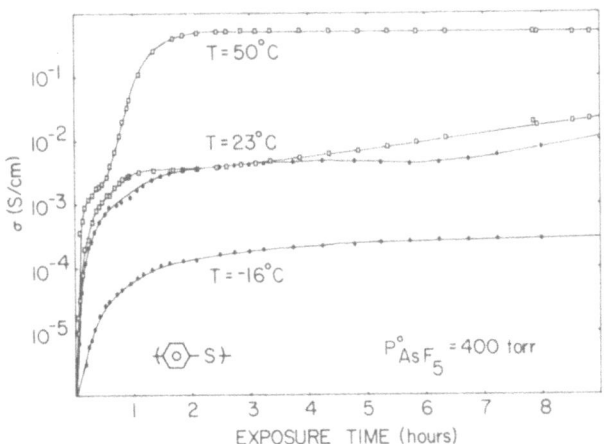

Fig. 6.   Conductivity of PPS pressed pellets versus time of exposure to AsF$_5$ at an initial pressure of 400 torr. The open squares (closed circles) represent the 50°C (-16°C) doping experiment with a control sample at 23°C exposed to the same gas.

TABLE 1:   VARIABLE TEMPERATURE AsF$_5$ DOPING OF PPS

Doping Temp. (°C)	σ_{max} (S/cm)	ΔE (eV)[a]	$[AsF_5]$[b]	Crosslinking[c]
−16°C	0.0035	0.16	0.13	None
+23°C	0.26	0.09	1.06	Substantial
+23°C	0.16	0.08	0.46	None
+50°C	0.60	0.08	0.69	Substantial

[a]Activation energy for σ in vicinity of room temperature

[b]Molar concentration of AsF$_5$ derived from weight increase.

[c]As judged from infrared analysis of compensated and washed samples.

−16°C doping experiments were carried out with a room temperature (23°C) control sample exposed to the same AsF$_5$ gas. The kinetics of doping are strongly temperature dependent. (This result is probably attributable to the temperature dependent dopant diffusion). However, from Table I no obvious differential in the temperature dependence of the doping and the crosslinking (judged from IR analysis) can be discerned in these data. The results do show that moderately high conductivities ($\sigma \sim 0.16$ S/cm) can be achieved without chemical modification.

SUMMARY

Though conductivities of doped PPS are lower than some of the other conducting polymers, the PPS system remains a particularly interesting system because of processibility considerations. We have also found that poly(m-phenylene sulfide) and poly(p-phenylene oxide) can be doped with AsF$_5$ to yield $\sigma \sim 10^{-2} - 10^{-3}$ S/cm. Thus, PPS is the first of an important new family of organic conductors involving phenyl chalcogen polymers and copolymers.

REFERENCES

1. Y. W. Park, M. A. Druy, C. K. Chiang, A. G. MacDiarmid,
    A. J. Heeger, H. Shirakawa, and S. Ikeda, J. Polym.
    Sci., Polym. Lett. Ed. 17, 195(1979) and references
    therein.
2. K. K. Kanazawa, A. F. Diaz, R. H. Geiss, W. D. Gill,
    J. F. Kwak, J. A. Logan, J. F. Rabolt, and G. B. Street,
    J.C.S. Chem. Comm. 854(1979); A. F. Diaz, K. K. Kanazawa,
    and G. P. Gardini, J.C.S. Chem. Comm. 635(1979).
3. L. W. Shacklette, R. R. Chance, D. M. Ivory, G. G. Miller,
    and R. H. Baughman, Synthetic Metals 1, 307(1980);
    D.M. Ivory, G. G. Miller, J. M. Sowa, L. W. Shacklette,
    R. R. Chance, and R. H. Baughman, J. Chem. Phys. 71,
    1506(1979).
4. R. H. Baughman, L. W. Shacklette, R. R. Chance, D. M. Ivory,
    G. G. Miller, A. F. Preziosi, and M. Lahav, Bull. Amer.
    Phys. Soc. 25, 314(1980).
5. R. R. Chance, L. W. Shacklette, G. G. Miller, D. M. Ivory,
    J. M. Sowa, R. L. Elsenbaumer, and R. H. Baughman,
    J. Chem. Soc. Chem. Comm. 348(1980).
6. J. F. Rabolt, T. C. Clarke, K. K. Kanazawa, J. R. Reynolds
    and G. B. Street, J. Chem. Soc. Chem. Comm. 347(1980).
7. N. F. Mott, "Metal Insulator Transitions" (Taylor and
    Francis, London, 1974) pp. 30-42.
8. E. K. Sichel, J. I. Gittleman, and Ping Sheng, Phys.
    Rev. B 18, 5712(1978) and references therein.
9. J. Boon and E. P. Magré, Makro. Chemie 126, 130(1969):
    B. J. Tabor, E. P. Magré and J. Boon, Europ. Polym.
    J. 7, 1127(1971).

MACROMOLECULAR METALS AND SEMICONDUCTORS: A COMPARATIVE STUDY

R.H. Baughman, J.L. Brédas,[*] R.R. Chance,
H. Eckhardt, R.L. Elsenbaumer, D.M. Ivory,
G.G. Miller, A.F. Preiziosi, and L.W. Shacklette

Corporate Research Center
Allied Chemical Corporation
Morristown, New Jersey 07960

INTRODUCTION

Highly conducting organic polymers represent a rapidly expanding new research area. As such, fundamental aspects of their electronic and structural properties are not well established and theoretical understanding lags far behind the pace of experimental discoveries.

The highly conducting organic polymers of interest here are metallic and semiconducting materials obtained by the addition of either electron donors or acceptors to essentially insulating precursor polymers. Via such doping it is possible to vary the electrical conductivity of poly(p-phenylene) in a controllable fashion over seventeen orders of magnitude – from less than 10^{-15} S/cm to over 500 S/cm.[1-3] Even higher conductivities (greater than 2000 S/cm), but a smaller conductivity range, has been obtained for AsF_5-doped, partially chain-oriented polyacetylene.[4]

This paper will provide an overview of research status. Emphasis will be on new conducting polymers, the structural nature of doped and undoped polymers, and electronic property variations which arise from structural differences.

THREE FAMILIES OF HIGHLY CONDUCTING ORGANIC POLYMERS

Polymers now known to form highly conducting compositions upon doping can be divided into three families depending upon the backbone type: polyenes,[5-8] poly(phenylenes),[1-3,9] and poly(phenylene chalcogenides).[10-11] These polymers are indicated in Tables I and II.

[*]Aspirant of the Belgian National Science Foundation (FNR)

Table I. Polyacetylene and Analogues: Conductivity
of Unoriented, Doped Polymers

Initial Polymer	Dopant	Dopant * Concentration	Conductivity (S/cm)	Ref.
$HC=CH$ / $C=C$ (H, H)	I_2	0.10	550	5
	$As F_5$	0.10	1200	5
(pyrrole ring, N–H)	(BF_4^-)	0.08	100	6
(thiophene ring, S)	I_2	0.03	3.4×10^{-4}	7
(cyclohexene ring structure)	I_2	~0.1	~0.1	8
$C=C\,C=C$ (H, H, H)	I_2	0.07	160	5
	$As F_5$	0.10	400	5

*Moles of triiodide species, BF_4^-, or arsenic fluoride species per $-CH=$
unit in backbone, neglecting the possibility of other dopant states.

Table II. Phenylene Polymers: Conductivity
of Unoriented, Doped Materials

Initial Polymer	As F_5/ Monomer	Conductivity (S/cm)	Ref.
(para-phenylene)	0.4	500	1 – 3
(meta-phenylene)	1.0	10^{-3}	
(phenylene–CH=CH–)	0.75	3	9
(phenylene–S–)	~1	1	10, 11
(phenylene–S—)	1.0	10^{-2}	10, 11
(phenylene–O–)	0.4	10^{-3}	10

Shirakawa and coworkers[12] reported in 1977 that polyacetylene
forms highly conducting complexes. The common sentiment was that
an essentialy unlimited category of conductors could be obtained
by replacing the hydrogens in polyacetylene by one or more substituent groups R, such as hydrocarbon sidechains. This expecta-
tion has not yet been realized. No polymer of the form
$(-(R)C=C(H)-)_x$ or $(-(R)C=C(R)-)_x$, with R other than hydrogen, has
been reported to dope to a high level of conductivity. Copolymers
of acetylene and substituted acetylenes (or homopolymer mixtures)
have been observed to dope to respectable conductivity levels.[13-14]
However, the reported conductivities are lower than obtained by
doping unsubstituted polyacetylene, suggesting the possibility
that the conductivity is principally due to lengths of unsubstituted polymer.

The absence of positive results for such substituted polyacetylenes might reflect increased ionization potentials, de-
stabilization of the planar backbone geometry due to steric interactions, decreased interchain interactions, or the low molecular
weights of investigated materials. Positive results obtained for
"chain-bridged polyenes" suggest that steric factors might be
important. Chain-bridged polyenes (polyacetylenes) are formed by
linking together the first and fourth carbon atoms in a C=C-C=C
chain segment or the first and third atom in a C=C-C chain segment
to form a five or six membered ring, as shown in Table I. Examples
of the first case are provided by poly(2,5-pyrrole)[6] and
poly(2,5-thienylene)[7] - which utilize NH and S, respectively, as
the chain-bridging elements. An example of the second class is
provided by poly(1,6-heptadiyne)[8], wherein $CH_2-CH_2-CH_2$ completes
a six membered ring with a C=C-C chain segment. Polymer structures
have not been determined. However, molecular models suggest that
these chain-bridging substituents can provide backbone structures
which are at least approximately planar.

We have found that highly conducting polymers can be obtained
by doping poly(p-phenylene)[1-3] and poly(m-phenylene) with electron
donors (alkalai metals) or electron acceptors (such as IF_5, SO_3,
HSO_3F, $SbCl_5$, and AsF_5 for the para polymer). Both these polymers
have a much higher ionization energy and optical bandgap than does
polyacetylene. In addition, the backbone structures of the undoped polymers deviate from planarity. Nevertheless, upon heavy
doping with AsF_5 both poly(p-phenylene) and polyacetylene evidence
similar free electron-like absorption spectra in the infrared,
similar d.c. conductivities, and similar thermopower values.[1-3,15]
The similarity is surprising if the properties of heavily doped
polyacetylene depend upon the existence of π-shift solitons,[16]
since the absence of two energetically equivalent backbone
structures for poly(p-phenylene) precludes the existence of a
strictly analogous boundary.

Poly(m-phenylene) differs from the abovementioned polyenes
and poly(p-phenylene) in that this polymer is both fusible and
soluble - permitting melt and solution processing. This results

from the greater flexibility of the poly(m-phenylene) backbone,
which is a consequence of the more localized electronic structure
of this polymer, as well as the increased configurational space
which is accessible for the meta-polymer. Possibly again re-
flecting the more localized electronic structure for poly(m-phenylene), the observed conductivities for unoriented polymer
(10^{-2} to 10^{-3} S/cm for AsF_5 doping) are much lower than
those of poly(p-phenylene).

Using the relationship of Berlin and Promyslov[17] and measured
ionization energies and optical transition energies of phenylene
oligomers and polymers, we have calculated the ionization
energies of poly(p-phenylene), 5.4 eV, and poly(m-phenylene),
6.5 eV. The calculated energy for the para polymer is in good
agreement with the ionization energy derived by Shacklette and
coworkers[18] from junction characteristics (5.4 eV). Both energies
are much higher than Salaneck et al.[19] reported for trans polyacetylene (4.7 eV). Consequently, it is not surprising that dopants
such as iodine form highly conducting complexes with polyacetylene,
but not with either polyphenylene isomer.

Since polyacetylene and poly(p-phenylene) become highly
conducting as a consequence of doping, one might anticipate high
conductivities for doped copolymers of the form
$[-(p-C_6H_4)_n(C=C)_m]_x$. Wnek et al[9] have shown that this is true
for poly(p-phenylene vinylene), where n and m are unity.

The third class of polymers which dope to high conductivities
is the most surprising of all – the phenylene chalcogenide
polymers (Table II). We have found[10] that poly(p-phenylene
sulfide), poly(m-phenylene sulfide), and poly(p-phenylene oxide)
form conducting complexes with AsF_5, a result which was also
obtained by Rabolt et al[11] for $(C_6H_4S)_x$. These melt and
solution processible parent polymers differ from those discussed
earlier in that there exists no continuous carbon atom chain in
the polymer backbone. In addition, the isomorphous poly(p-phenylene
sulfide) and poly(p-phenylene oxide) polymers are grossly
nonplanar.[20,21]

STRUCTURAL ASPECTS RELEVANT FOR ELECTRONIC PROPERTIES

Because all of the investigated polymers are insoluble and
infusible after doping with AsF_5, it is difficult to establish
fundamental structural properties. The problem is further complicated by the polycrystalline or amorphous nature of these
materials, the introduction of additional disorder upon doping,
free electron absorption which obscures molecular vibrations, and
possible heterogeneity of the doped materials.

We have determined the crystal structures of cis and trans
polyacetylene and poly(p-phenylene) using a variety of indirect
methods. These include energy minimization packing calculations,
structure prediction with model compounds, and analysis of the
limited amount of available X-ray diffraction data.[22,23] It is

interesting that both isomeric forms of polyacetylene and
poly(p-phenylene) evidence similar crystal packing arrangements.
Despite the likely monoclinic symmetry of trans-polyacetylene,
$P2_1/a$, and poly(p-phenylene), $P2_1/a$, and the orthorhombic
symmetry of cis-polyacetylene, Pnma, all of these polymers display
the same symmetry in chain axis projection (Pgg). In each case
a herringbone-like packing arrangement is obtained, which provides
little overlap between parallel π-orbitals in neighboring
molecules.[2,3,22,23]

There is presently no evidence which suggests that cis or
trans polyacetylene chains deviate from planarity. The quantum
mechanical calculations of Brédas et al.[24] indicate an energy
minima for the planar configuration. However, due to the steric
repulsion between ortho hydrogens in poly(phenylene), a non-planar
backbone structure arises. Extrapolating from structural results
for phenylene oligomers,[2,3,23,25] the equilibrium low temperature
phase of poly(p-phenylene) should have an ordered, nonplanar
structure with approximately a 23° angle between adjacent phenyls.
Oppositely directed rotations for adjacent phenyl rings would
provide a two-monomer-long chain repeat length. At about 450°C,
a second-order or a pseudo second order phase transition is predicted from oligomer data - at which point the polymer chains lose
long-range order in the relative rotation of phenyls.[23]

A more extreme case is provided for the poly(phenylene chalcogenides). The chalcogenide atoms in poly(p-phenylene sulfide)
and poly(p-phenylene oxide) form a planar zig-zag chain.[20,21]
However, neighboring phenyl rings along the chain form an angle,
with respect to this plane, of +40° and -40° for the oxide[20] and
+45° and -45° for the sulfide.[21] It is interesting that much
higher conductivities are observed for the AsF_5-doped sulfide than
for the oxide. The difference in conductivity may result from
a lower ionization potential for the sulfide than the oxide.
Duke and Paton[26] suggest, based on oligomer calculations, that
the p-orbitals of sulfur and oxygen provide an extended
π-systems in the poly(p-phenylene sulfide) and poly(p-phenylene
oxide) polymers, which is consistent with the strong chain
axis polarization which we observe for the π-π* transition
of poly(p-phenylene sulfide).

As bond-length alternation decreases to zero in trans-
polyacetylene, the optical bandgap is expected to vanish -
providing a metallic conductor. However, at present there is
no experimental evidence which proves that doping polyacetylene
removes bond-alternation in this polymer. Poly(p-phenylene)
provides a much more localized electronic structure than does
polyacetylene. The bond connecting adjacent phenyl rings is
extremely long (1.50Å, using phenyl oligomers as model compounds), compared with the 1.40 - 1.41Å bond lengths in the phenyl
rings, and the 1.46Å "single bond" in polyacetylene.[22,25] If
alkalai metal-doped poly(p-phenylene) behaves analogously to
alkalai metal salts of the biphenyl anion, $C_6H_5-C_6H_5^-$, [23-27]

the C-C bonds parallel to the chain direction would shorten
and those inclined to the chain direction would elongate during
doping - but adjacent phenyl rings would remain slightly non-
planar. This bond change amounts in biphenyl to about -0.04Å
for the phenyl-phenyl carbon bond. These changes bring all
C-C bonds to within a 0.05Å difference, which is about
one-half the range of C-C bond lengths in neat biphenyl
crystals and in undoped polyacetylene.[22,23,27]
     There is an apparent correlation of electrical conductivity
with homogeneity of the polymer backbone. The regular copolymers
in Table III dope to lower conductivity levels than do homopolymers containing any one of the constituent chain elements.
This result is not unreasonable, since chemical heterogeneity
along a polymer backbone can provide carrier localization on the
chain unit which provides the lowest potential for holes (or
electrons in the case of donor doping). Exceptions to this
correlation might be expected in cases where copolymerization
removes steric hindrances to planarity which are present in
one of the homopolymers or where one of the chain elements
interrupts electronic connectivity of the polymer backbone.
Nevertheless, the data in Table III does give some guidance
which can be used in synthetic efforts. To obtain polymers
having the highest conductivities, it would appear that one
should concentrate on polymers having the most homogeneous
chain structures.
     It is found that the introduction of a methylene linkage
provides a sufficient barrier to dramatically decrease electrical
conductivity for the doped polymers. For example, the replacement

Table III.   The Conductivity of AsF$_5$ Doped
             Polymers and Derivative Copolymers[a]

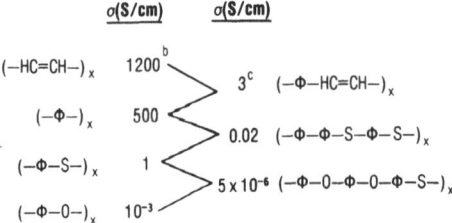

[a] The effect of differing morphologies, molecular weights and doping
    kinetics is unknown.
[b] Ref. 5
[c] Ref. 9

of every other chain chalcogen with a methylene in poly(p-phenylene sulfide) or poly(p-phenylene oxide) reduces the electrical conductivity for the AsF_5 doped polymer from 1 S/cm and 10^{-3} S/cm, respectively, to less than 10^{-6} S/cm. Likewise the insertion of methylenes between the phenyls of poly(p-phenylene) provides a polymer which does not dope to high conductivity.

The dependence of electrical conductivity on the molecular weight of doped polymers is not well established. AsF_5 doping of phenylene oligomers provides a molecular weight increase via para coupling of adjacent chain ends, which precludes ready correlation of molecular weight and conductivity for the phenylene polymer/AsF_5 system.[3,28] The poly(p-phenylene) prepared by the Kovacic method contains 8 to 16 phenyls on the average, corresponding to an average chain length of no more than 70Å. End-group analysis indicates that the chain length can more than double upon AsF_5 doping which provides a conductivity greater than 500 S/cm.[3] Low degrees of polymerization are indicated for the poly(2,5-thienylene),[7] poly(m-phenylene),[31] and poly(p-phenylene vinylene)[9] polymers which dope to the conductivity levels indicated in Tables I and II. The possibility of crosslinking or chain extension caused by AsF_5 doping is not excluded for these polymers. In fact, AsF_5 doped and dimethylamine compensated poly(m-phenylene) and poly(p-phenylene sulfide)[10,11,32,33] are both insoluble and infusible - in contrast with the parent polymers. Furthermore, Wnek et al.[9,34] find HF evolution during the AsF_5 doping of poly(p-phenylene vinylene) and a conductivity independent of the degree of polymerization before doping, which is consistent with chain growth during doping. Permanent chemical modification does not occur for alkalai metal doped p-phenylene oligomers, where the parent compound can be recovered from the complex.[28] Conductivities obtained for potassium doped sexi-phenyl, $H_5C_6(C_6H_4)_4C_6H_5$, are about 0.5 S/cm as compared with about 30 S/cm for the polymer prepared by the Kovacic method.[28] Polyacetylene appears to provide quite long chain lengths, estimated to be between about 600 Å and 5000Å.[35-37] However, because of conformational and/or chemical defects either in the undoped polymer or arising from the doping process, it is unlikely that the effective conjugation length is nearly so long.

X-ray diffraction results on acceptor doped polyacetylene suggest the formation of layer structures analogous to those for intercalated graphite.[22,38] Long X-ray diffraction spacings are observed for the doped polymer which are not present for the parent polymer. As shown in Fig. 1, these long X-ray diffraction spacings increase in proportion to the effective van der Waals thickness of the dopant, as measured by the amount graphite would expand for the same dopant. Hence, the long X-ray diffraction spacings can be associated with the distance between planes of polyacetylene chains separated by a layer of dopant molecules. NMR second moment measurements as a function of iodine doping by Mihály et al.[39] are consistent with such an intercalated structure.

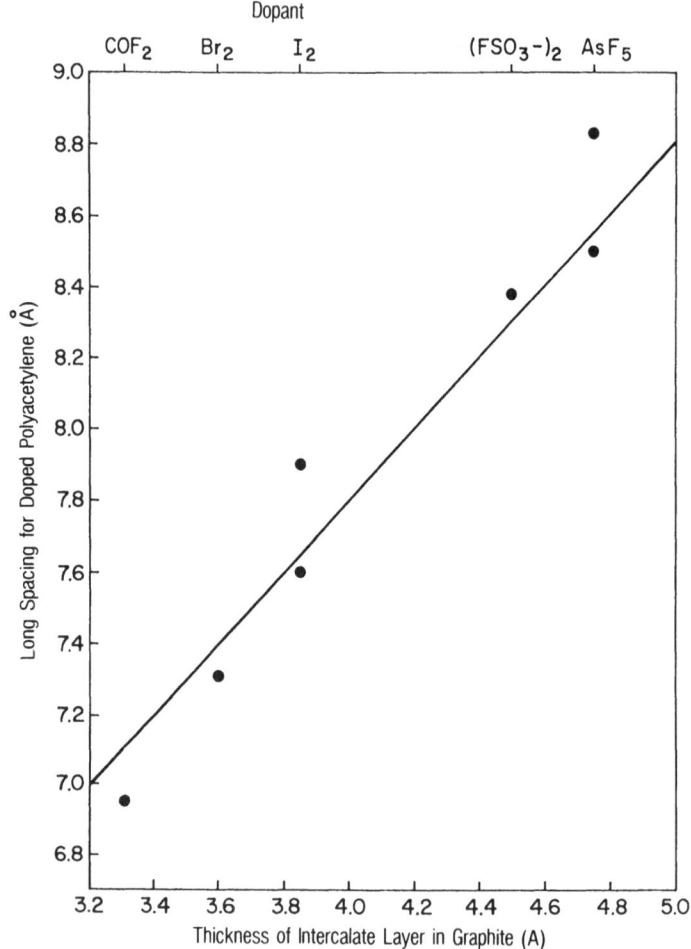

Figure 1. The long interplanar spacing for acceptor doped polyacetylene as a function of the effective thickness of a dopant
layer in graphite.[22,38] The iodine/graphite result is derived from measurements for bromine/graphite.

Thus far, there is no evidence indicating intercalation structures
for other conducting polymers, or for donor-doped polyacetylene.
Moreover, at lower acceptor dopant levels the picture is
complicated by the possibility of nonuniform doping, higher
stage layer structures, and even nonlayered structures.

Molecular packing calculations for polyacetylene and
poly(p-phenylene) suggest that the most likely layer intercalation
structure would involve interchain hydrocarbon interactions
similar to those in alternate-stack molecular charge-transfer
crystals. Consequently, it is possible that high conductivities
will result both in the chain direction and in a direction inclined with respect to the chain axis. Such bidirectional conductivity might explain the relatively low electrical anisotropy
observed for chain oriented poly(p-phenylene) and poly(p-phenylene
sulfide) as well as the high bulk conductivities observed for
low molecular weight, unoriented doped-polymers.

Limited information is available on the general effect of
order on the conductivity of the doped polymers. Deitz and
coworkers [40] have shown that the electrical conductivity of
iodine-doped polyacetylene strongly depends upon the polymer
synthesis method. They observed about a six order of magnitude
difference in electrical conductivity for polymerized acetylene
samples having comparable dopant levels ($CHI_{0.15-0.25}$) and long
initial conjugation lengths, evidenced by the UV-visible
spectroscopy of the undoped polymer. Crystallinity differences
between samples appeared to be critical for explaining the
widely differing conductivities. All crystalline samples doped
to high conductivities, all amorphous samples provide poorly
conducting materials. However, differences in the chemical
structure of the chain (crosslinking, oxygen impurity, etc.) and
morphology cannot be excluded as important determinants for
conductivity in the doped polymers. Perhaps the correlation
with degree of order is secondary in the sense that chemical
structure inhomogeneities might provide the low degrees of order
for the undoped polymers. Even if the chemical structures of the
undoped polymers are essentially identical, this need not be true
for the doped polymers. More specifically, MacDiarmid and
coworkers[41] have shown that iodine reacts slowly with well ordered
polyacetylene to produce gaseous HI. Such reaction is likely to
occur more readily in amorphous polymer than in crystalline
polymer, so the structure-conductivity relationship could be a
consequence of decreased conjugation length via partial iodination
of the amorphous polymer.

No significant three-dimensional order is apparent for
AsF$_5$-doped poly(p-phenylene)[3] and poly(p-phenylene sulfide)[10,11]
despite the high conductivities of these materials. Major
conductivity differences are observed upon AsF$_5$ doping of
poly(p-phenylene) prepared using different synthesis methods,
but no obvious correlation is observed between the conductivity

and the crystallinity of the precursor undoped polymer. Likewise, comparable conductivity levels are obtained for the AsF_5 doping of either amorphous or crystalline poly(p-phenylene sulfide). Finally, high conductivities are reported for iodine doping of amorphous poly(1,6-heptadiyne).[8] Three dimensional order either in the doped polymers or in the precursor undoped polymers is clearly not a general prerequisite for high conductivity levels.

SUMMARY

Despite progress in discovering a variety of new metallic and semiconducting polymeric conductors, the structure basis for predicting electronic properties or identifying new conductors is quite limited. Fundamental uncertainties remain for all systems regarding the dopant species, effect of chain ends and dopant-induced reactions, polymer-dopant packing, and the inherent anisotropic electronic properties. The solution to these problems will likely require continued investigation of structural and property aspects of monocrystalline model compounds, as well as continued success in eliminating disorder in parent and doped polymers.

REFERENCES

1.   D.M. Ivory, G.G. Miller, J.M. Sowa, L.W. Shacklette, R.R. Chance, and R.H. Baughman, J. Chem. Phys. 71, 1506 (1979).

2.   R.H. Baughman, D.M. Ivory, G.G. Miller, L.W. Shacklette, and R.R. Chance, Organic Coatings and Plastics Chemistry 41, 139 (1979).

3.   L.W. Shacklette, R.R. Chance, D.M. Ivory, G.G. Miller, and R.H. Baughman, Synthetic Metals, 1, 307 (1980) and references therein.

4.   Y.W. Park, M.A. Druy, C.K. Chiang, A.G. MacDiarmid, A.J. Heeger, H. Shirakawa, and S. Ikeda, J. Polym. Sci., Polym. Lett. Ed. 17, 195 (19790.

5.   A.G. MacDiarmid and A.J. Heeger, Synthetic Metals 1, 101 (1980).

6.   K.K. Kanazawa, A.F. Diaz, R.H. Geiss, W.D. Gill, J.F. Kwak, J.A. Logan, J.F. Rabolt, and G.B. Street, J.C.S. Chem. Comm. 854, (1979).

7.   T. Yamamoto, K. Sanechika, and A. Yamamoto, J. Polym. Sci., Polymer Lett. Ed. 18, 9 (1980).

8.   H.W. Gibson, F.C. Bailey, J.M. Pochan, A.J. Epstein, and H. Rommelmann, Organic Coatings and Plastics Chemistry 42, 603 (1980).

9.   G.E. Wnek, J.C.W. Chien, F.E. Karasz, and C.P. Lillya, Polymer 20, 1441 (1979).

10. R.R. Chance, L.W. Shacklette, G.G. Miller, D.M. Ivory, J.M. Sowa, R.L. Elsenbaumer, and R.H. Baughman, J.C.S. Chem. Comm. 347 (1980).

11. J.F. Rabolt, T.C. Clarke, K.K. Kanazawa, J.R. Reynolds, and G.B. Street, J.C.S. Chem. Comm. 348 (1980).

12. H. Shirakawa, E.J. Louis, A.G. MacDiarmid, C.K. Chiang, and A.J. Heeger, J.C.S. Chem. Comm. 578 (1977).

13. M.J. Kletter, T. Woerner, A. Pron, A.G. MacDiarmid, A.J. Heeger, and Y.W. Park, J.C.S. Chem. Comm., in press.

14. G.E. Wnek, J.C.W. Chien, and F.E. Karasz, Organic Coatings and Plastics Chemistry 43, 882 (1980).

15. C.R. Fincher, Jr., M. Ozaki, M. Tanaka, D. Peebles, L. Lauchlan, and A.J. Heeger, Phys. Rev. B 20, 1589 (1979).

16. B.P. Weinberger, J. Kaufer, A.J. Heeger, A. Pron, A.G. MacDiarmid, Phys. Rev. B 20, 223 (1979).

17. A.A. Berlin and V.M. Promyslov, Zhurnal Strukturnoi Khimii 11, 1076 (1970).

18. L.W. Shacklette and coworkers, unpublished.

19. W.R. Salaneck, H.R. Thomas, C.B. Duke, A. Paton, E.W. Plummer, A.J. Heeger, and A.G. MacDiarmid, J. Chem. Phys. 71, 2044 (1979).

20. J. Boon and E.P. Magré, Makromolekulare Chemie 126, 130 (1969).

21. B.J. Tabor, E.P. Magré and J. Boon, Europ. Polym. J. 7, 1127 (1971).

22. R.H. Baughman, S.L. Hsu, L.R. Anderson, G.P. Pez, and A.J. Signorelli, Molecular Metals, NATO Conference Series, W.E. Hatfield Ed. (Plenum Press, 1979), p. 187 and references therein.

23. R.H. Baughman et al., unpublished.

24. J.L. Brédas, Ph.D. Thesis, Facultés Universitaires de Namur, Belgium, 1979.

25. Y. Delugeard, J. Desuche, and J.L. Baudour, Acta. Cryst. B32, 702 (1976).

26. C.B. Duke and A. Paton, Organic Coatings and Plastics Chemistry 43, 863 (1980).

27. E. de Boer, A.A.K. Klaasen, J.J. Mooij and J.H. Noordik, Pure and Appl. Chem. 51, 73 (1979) and references therein.

28. L.W. Shacklette, H. Eckhardt, R.R. Chance, G.G. Miller, D.M. Ivory, and R.H. Baughman, J. Chem. Phys., in press.

29. J.E. Durham and P. Kovacic, J. Polym. Sci. Polym. Chem. Ed., 15, 2701 (1977).

30. M.B. Jones, P. Kovacic, and D. Lanska, J. Polym. Sci. Polym. Chem. Ed., in press.

31. A.F. Preziosi and coworkers, unpublished; T. Yamamoto, Y. Hayashi, and A. Yamamoto, Bull. Chem. Soc. Japan 51, 2091 (1978).

32. R.R. Chance, L.W. Shacklette, H. Eckhardt, J.M. Sowa,
     R.L. Elsenbaumer, D.M. Ivory, G.G. Miller, and
     R.H. Baughman, Organic Coatings and Polymer
     Chemistry 43, 768 (1980).
33. J.F. Rabolt, T.C. Clarke, KK. Kanazawa, J.R. Reynolds,
     and G.B. Street, Organic Coatings and Polymer
     Chemistry 43, 772 (1980).
34. J.C.W. Chien, R.D. Gooding, F.E. Karasz, C.P. Lillya,
     G.E. Wynek, and K. Yao, Organic Coatings and Plastics
     Chemistry 43, 886 (1980).
35. H. Shirakawa, M. Sato, A. Hamano, S. Kawakami, K. Soga,
     and S. Ikeda, Macromolecules 13, 459 (1980).
36. J.C.W. Chien, J.D. Capistran, L.C. Dickinson, and
     F.E. Karasz, Organic Coatings and Plastics Chemistry
     43, 875 (1980).
37. G. Lieser, G. Wegner, W. Müller, and V. Enkelmann, preprint.
38. T.C. Clark, R.H. Geiss, W.D. Gill, P.M. Grant, H. Morawitz,
     G.B. Street, and D.E. Sayers, Synthetic Metals 1, 21
     (1980).
39. L. Mihály, S. Pekker, and A. Jánossy, Synthetic Metals 1,
     349 (1980).
40. W. Deits, P. Cukor, and M. Rubner, Organic Coatings and
     Plastics Chemistry 43, 867 (1980).
41. A.G. MacDiarmid and coworkers, unpublished.

ELECTROSYNTHESIS AND STUDY OF CONDUCTING POLYMERIC FILMS

A. F. Diaz, K. K. Kanazawa, J. I. Castillo, and J. A. Logan

IBM Research Laboratory
San Jose, California 95193

This report briefly describes the electrochemical preparation[1] and behavior[2] of the polypyrrole films. The orientation of this report was selected because a more detailed overview of the properties of these films was recently reported in the proceedings to the International Conference on Low Dimensional Synthetic Metals.[3]

With polypyrrole we have demonstrated the attractiveness of the stoichiometric electropolymerization reaction as an approach for preparing polymers with electroactive properties and variable conductivity. The electropolymerization of pyrrole produces thin films of polypyrrole on the surface of the anode. The films consist of linear polymers of pyrrole which are generated in the oxidized form. The elemental analyses of thick (20-50 μm) free standing films indicate that they are 70% (by wt) polymer and 30% anion (ideally BF_4^-, the anion of electrolyte). The anion of the electrolyte is affiliated by the cationic polymer and has a stoichiometry of one anion/four pyrrole rings. Not only was this stoichiometry determined from the elemental analyses but it was corroborated by the coulometric studies of the film preparation where the number of Faradays per mole can be determined. This point is discussed in more detail below. These thick films when in the oxidized form, display metal-like conductivity, have good electrical contact to the platinum electrode surface (where they were generated) and are stable in air and in most solutions.

The electrosynthesis of these films proceeds via the oxidation of pyrrole at the platinum anode (E_{pa} equals +1.2V versus sodium chloride calomel electrode (SSCE)) to produce an unstable π-radical cation (I) which then reacts with the neighboring pyrrole species (Eq. (1)). The mechanism of the overall reaction for the formation of the fully aromatized product is very complicated and involves a series of oxidation and deprotonation steps (Eq. (2)). It is not clear at this time whether the coupling reaction involves the coupling of two pyrrole radical cations or of a neutral and a radical cation species. This reaction has electrochemical stoichiometry where two electrons/pyrrole ring are involved in the formation of the polymer. We in fact measure 2.2-2.4 electrons/pyrrole ring, however, the extra charge (0.2-0.4) consumed during the reaction results from the concurrent oxidation of the resulting polymer which has a lower oxidation potential ($E° = -0.2V$) than the monomer. This electrochemical stoichiometry of this reaction precludes the presence of a polymerization mechanism involving a cationic propagation step. In this regard, this reaction is different from the electrochemical polymerization reactions normally found in the literature where the reaction is initiated at the electrode surface but the polymer is formed in a chain propagation step which occurs in the bulk of the solution. Because the film remains on the electrode surface, the conducting nature of the film is important for the continuation of the reaction. The stoichiometry of this reaction also provides a real convenience in the preparation of these films since the desired thicknesses can be controlled by monitoring the current density of the reaction.

$$(1)$$

$$(2)$$

$$(3)$$

Thin films of these materials (ca. 0.1 μm) are electroactive and when left attached to the electrode can be driven repeatedly between the oxidized (conducting) and neutral (nonconducting) state. This reaction was studied by cyclic voltammetry where the potential applied to the working electrode was varied linearly from -0.5V to 0.6V to -1.2V and back to -0.5V. In this cycle, a wave for the oxidation of the polymer appears at -0.1V (versus SSCE) in the anodic sweep while the wave for the reduction reaction appears at -0.3V in the cathodic sweep. The integrated charges under the oxidation and the reduction waves are equal and amount to ca. 0.3 electron charges per pyrrole unit. This value is again in accord with the charge concentration in the polymer determined from the anion affiliation discussed earlier. The shape of the oxidation and reduction waves are different suggesting that the kinetics of the two reactions are different. This difference must not be due to the polymer alone since it is not involved in a diffusion process during this reaction, but instead remains localized on the electrode surface. Furthermore, the wave shape is found to depend on the nature of the anion in the electrolyte suggesting that the kinetics of the electron transfer process is primarily limited by the diffusion of the counter ion in the film.

This redox reaction involves electron transfer in and out of the extended π-system of the polymer (Eq. (3)). The film is stable to this reaction and can be driven for extended periods of time with no evidence of decomposition. Because the electrical contact is good between the film and the electrode, the redox reaction is fast and occurs within milliseconds (0.1 μm film thickness).

The switching time is drastically increased as the thickness of the film increases. In addition, it is increased in the presence of oxygen. Therefore, the thicker films of polypyrrole (0.5-2 μm) perform very well as a working electrode in a conventional electrochemical cell. Unlike the thin films, these thicker films are very difficult to switch in practice and instead remain conducting even in the region cathodic of -0.2V. The electrode behavior of the films was probed using the redox reactions of compounds which are known to have reversible electrochemical behavior on a platinum electrode, for example, the reaction of ferrocene, chloranil and phenothiazine. The observed $E°$ values for these reactions were found to be the same as those measured on a platinum electrode and the heterogeneous electron transfer process is rapid. The films are stable in this application and can be used repeatedly.

Conducting polymer films are also produced using N-substituted pyrrole derivatives as long as the pyrrole ring is not substituted in the α-position. The pyrrole ring and the substituent remain intact in this reaction which provides us with a very convenient method for preparing derivatized conducting polymers. For example, a series of pyrrole polymers with N-alkyl groups (methyl, ethyl, propyl, butyl) were prepared.[2b] With these chemical modifications, some changes are observed in the properties of the films. The conductivity decreases by a factor of 10^5 and the degree of oxidation of the polymer decreases along the series, from 0.3 charges/pyrrole ring for polypyrrole to 0.1 charges/pyrrole ring for poly-N-butylpyrrole. The oxidation potential is also affected where the $E°$ values for the various N-alkyl substituted polypyrrole polymers are in the range 0.45V to 0.64V versus SSCE (measured on a platinum electrode in acetonitrile containing Et_4NBF_4). Within this series the larger alkyl groups produce the higher $E°$ values. In contrast with the electrochemical behavior of polypyrrole, however, the redox reaction of the various substituted polypyrroles is much less sensitive to the presence of oxygen.

Finally, copolymer films can be prepared by copolymerizing monomer mixtures, e.g., pyrrole and N-methylpyrrole.[1] The resulting films have conductivities and redox potentials which are intermediate between the values of the pure polymers.[2,3] In fact, the cyclic voltammograms show two sets of oxidation and reduction waves which may suggest that block polymer regions are present in the film.[3] Inspection of the relative peak areas in the voltammogram suggests that the pyrrole/N-methylpyrrole composition in the resulting polymer does not equal the composition of the monomeric mixture in the electrolytic solution. Instead there is a preference for the incorporation of pyrrole over the N-methylpyrrole by a factor of 2-3. We find that the films can also be prepared using a wide variety of electrolyte salts, and solvents for that matter, as long as the nucleophilicity of the anion-solvent combination does not compete with pyrrole for capture of the pyrrole radical cation intermediates. A variety of different films have been prepared where the anion is BF_4^-, ClO_4^-, AsF_6^-, halides, carboxylates and sulfonates. Because the anion makes up as much as 30% of the film by weight, it influences the properties of the film and we can use this effect to further vary the properties of the films. For example, the conductivity of polypyrrole films can be varied by a factor of 10^5 when the anion changes from BF_4^-, ClO_4^- or AsF_6^- to toluenesulfonate to oxylate anion, respectively.

EXPERIMENTAL

Reagents and solvents were purified in the usual manner. The solutions used for the preparation of the films contained 0.05M of the appropriate pyrrole derivative and 0.1M electrolyte salt. The solutions were purged briefly with helium in order to remove traces of oxygen. The solutions used for analyses of the films contained 0.1M electrolyte salt and in those cases where the solutions were used to test the electrode properties of the films, the solutions contained added 10^{-3}M ferrocene, nitrobenzene or chloranil.

Film Preparation. Platinum films (0.5 cm^2) deposited on glass plates were used as the working electrodes for the electropolymerizations. The film preparations were performed in a one compartment cell containing the appropriate electrolyte solution, the platinum film working electrode, a gold wire counter electrode and an aqueous sodium chloride calomel reference electrode. The thin films for electrochemical analyses were grown at constant potential using ca. 160 μA/cm^2 current density and in the absence of oxygen. Typically, the films of polypyrrole were prepared by oxidizing pyrrole (E_{pa}, +1200 mV) at 810 mV, those of poly-N-methylpyrrole by oxidizing N-methylpyrrole (E_{pa}, +1190 mV) at 800 mV and those of poly-N-phenylpyrrole (E_{pa}, +1800 mV) at +1200 mV. The film thickness was controlled by the amount of charge passed. Films prepared using 8 mC/cm^2, 24 mC/cm^2 and 32 mC/cm^2 were routinely employed. After the films were formed, the solution was removed from the cell and the cell was rinsed several times with fresh electrolyte solutions containing no pyrrole derivative and refilled again for analysis by cyclic voltammetry. Care was taken to keep oxygen out of the cell during these washings. All measurements were prepared with 60-80% IR compensation.

Thicker films (20-50μ) for elemental analysis and physical measurements were prepared using a larger area platinum electrode (7 cm^2) and at constant current density, ca. 500 μA/cm^2. The films were removed from the electrode surface, washed with acetonitrile in a Soxhlet extractor for one hour and dried in a vacuum. All the electrochemical equipment used in this study was designed and built in our laboratories.[4]

REFERENCES

1.   (a) A. F. Diaz, K. K. Kanazawa and G. P. Gardini, J. Chem. Soc., Chem. Comm., 635 (1979); (b) K. K. Kanazawa, A. F. Diaz, R. H. Geiss, W. D. Gill, J. F. Kwak, J. A. Logan, J. F. Rabolt and G. B. Street, J. Chem. Soc., Chem. Comm., 854 (1979); (c) K. K. Kanazawa, A. F. Diaz, W. D. Gill, P. M. Grant, G. B. Street, G. P. Gardini and J. F. Kwak, Syn. Metals 1, 329 (1980);

(d) K. K. Kanazawa, A. F. Diaz, M. T. Krounbi and G. B. Street, to be submitted.

2. (a) A. F. Diaz and J. I. Castillo, J. Chem. Soc., Chem. Comm., 397 (1980);
   (b) A. F. Diaz, J. Castillo, K. K. Kanazawa, J. A. Logan, M. Salmon and O. Fajardo, to be submitted.

3. A. F. Diaz, "Electrochemical Preparation and Characterization of Conducting Polymers," Proceedings to the International Conference on Low Dimensional Synthetic Metals, Chemica Scripta, 1981.

4. K. K. Kanazawa and R. Galwey, J. Electrochem. Soc. 124, 1385 (1977).

DEPENDENCE OF POLYMER ELECTRONIC STRUCTURE ON MOLECULAR
ARCHITECTURE: POLYACETYLENES, POLYPHENYLENES AND
POLYTHIENYLENE

C.B. Duke and A. Paton

Xerox Webster Research Center
Xerox Square-114, Rochester, N.Y., 14644

INTRODUCTION

Increasingly frequent attempts to prepare conducting polymeric materials via the molecular doping of nominally insulating nonsaturated-backbone polymer matrices has led to recent studies of
the electrical   properties of doped polyacetylene$[(CH)_x]$ (1-4),
poly(p-phenylene) $[(C_6H_4)_x]$ (5),   poly(2,5-thienylene) $[(C_4SH_2)_x]$
(6), poly(1,6-heptadiyne) $([CH(C_6H_7)]_x)$ (7), and poly(p-phenylene
sulfide) $[(C_6H_4-S)_x]$ (8,9). Consequently, the nature of the
electronic states in such materials is a matter of considerable
interest, and several different types of models have been proposed
to describe these states (10). In the case of the doped polymers,
the construction of suitable models for this purpose is hampered
severely by the lack of structural information on the positions of
the dopants. Even for "pure" polymers, however, the conformations
of the macromolecular chains are known precisely only in certain
special cases of crystalline (isotactic) materials. Therefore in
this paper we report a study of the effects of molecular architecture and conformation on the cation states of model oligomers.

We focus our attention on two issues. First, what is the
influence of the molecular architectures of the various polymer
chains on their high-energy π -electron valence states. Second,
what are the consequences of differing backbone conformations on
these states? We anticipate that the nature of these high-energy
cation (hole) states will play an important role in the electrical
properties of acceptor (e.g., halogen, AsF_5) doped polymers
because holes are regarded as being the mobile carriers in one-
electron models of these materials (10,11).

155

MODEL CALCULATIONS

All of our calculations were performed for electrically neutral molecules using the CNDO/S3 molecular orbital model (12). The molecular orbital eigenvalues of the neutral species are identified with the associated molecular cation energies in accordance with the procedures developed during the construction of this model. Since a review of the details and applications of the model recently has been given by Duke (12), we do not recapitulate them here. The parameters used in the calculations [as described in reference (12)] are specified in Table I. Oxygen as well as sulfur parameters are given in the table because we examine both poly(p-phenylene oxide) and poly(p-phenylene sulfide).

TABLE I. Parameters used to define the CNDO/S3 model. The interatomic Coulomb integrals are specified in terms of their intraatomic values (γ_A, γ_B) and the distance between the atomic centers (R_{AB}) via $\gamma_{AB} = 14.397 [28.794 (\gamma_A + \gamma_B)^{-1} + R_{AB}]^{-1}$.

Atom	I_s(eV)	I_p(eV)	B_s(eV)	B_p(eV)	γ(eV)	$c(A^{-1})$
H	13.60	——	10	—	12.85	2.33
$C(sp^2)$	21.34	11.54	20	17	10.63	3.78
$C(sp^3)$	21.34	11.54	20	17	10.63	3.07
O	35.50	17.91	31	26	13.10	4.32
S	21.02	10.97	18	15	9.67	4.37(s)
						3.80(p)

Comparison between the CNDO/S3 model calculations and measured photoemission spectra is achieved via plots of the density of valence states, i.e. "DOVS" (13). These plots are obtained by representing each CNDO/S3 eigenvalue by a normalized gaussian of width β and weight equal to twice the degeneracy of the eigenvalue. For typical gas-phase photoemission spectra 0.2 eV $\leq \beta \leq 0.3$ eV whereas for photoemission from polymers and molecular solids 0.4 eV $\leq \beta \leq 0.7$ eV (14-16). The large values of β for condensed molecular phases have the consequence that for the polymers considered herein, photoemission from oligomers becomes identical to that from macromolecules when the oligomers contain 16-20 carbon atoms in the backbone chain (10,16). Therefore it is not possible to verify via photoemission spectroscopy phase coherence within macromolecules over distances greater than about 25 A. A complete theory of the widths of photoemission lines from polymers and molecular solids has been constructed elsewhere

(17,18) and shown to describe both the magnitude (18) and temperature dependence (19) of observed spectra. This theory reveals that
the large values of $\beta \approx$ 0.6 eV for polymers are associated in part
with the photoemission process itself (17) and in part with the
disordered nature of the polymeric solid state (17,18). Hence,
from the perspective of photoemission spectroscopy, oligomers
containing 16 or more backbone carbon species are indistinguish-
able from macromolecules.

RESULTS

     The results of our calculations are orbital eigenvalues,
eigenfunctions, and DOVS as functionals of molecular architecture
and conformation. It is convenient to examine the nature of the
valence electron states in acetylene-related nonsaturated-backbone
polymers by considering these materials to be built up in a series
of steps. First, for polyacetylene, $(CH)_x$, the consequences of
molecular conformation on the π-electron valence orbitals are
analyzed. Next, the influence of the inclusion of aliphatic propyl
closed-ring side groups on the orbitals is assessed via the study
of oligomers of poly(1,6-heptadiyne), i.e., $[CH(C_6H_7)]_x$. The
effects of the incorporation of benzene aromatic groups into the
backbone are illustrated by the oligomers of poly(p-phenylene),
i.e., $(C_6H_4)_x$. The inclusion of an asymmetric six-π-electron
ring is achieved by replacing the benzene ring by a thiophene ring
thereby generating poly(1,6-thienyl), or $(C_4SH_2)_x$. The consequences of linking benzene moieties in the backbone via intermediate oxygen, and sulfur groups are indicated by calculations for
oligomers of poly(p-phenylene oxide), $(C_6H_4-O)_x$, and poly(p-phenylene sulfide), $(C_6H_4-S)_x$. Experimental molecular geometries are
used where available. Otherwise "standard" bond lengths and bond
angles (20) are utilized, e.g. $< CCC = 120$, $d_S = 1.35$ A and $d_L =$
1.44 A for the bond angles, short (S) and long (L) bond lengths in
the polyacetylene chain.

Polyacetylene

     One of our major interests in polyacetylene is the influence
on the DOVS of the conformation of the $(CH)_x$ chain. Considering
first the simplest representative model molecule, we show in Fig. 1
the CNDO/S3 DOVS for $H(CH)_8H$ in each of its four possible planar
sp^2 conformations. The highest-energy three orbitals are the
three π orbitals bonding across the short bonds and exhibiting
3,2 and 1 nodes, respectively, across the long bonds. The energies
of these sp^2 orbitals are essentially invariant under changes of
planar sp^2 conformation, although the energies of the lower-energy
sigma orbitals vary quite noticeably under these changes. A
similar result is obtained for $H(CH)_{16}H$ as shown in Fig. 2. In
both cases, differences between the various trans conformations

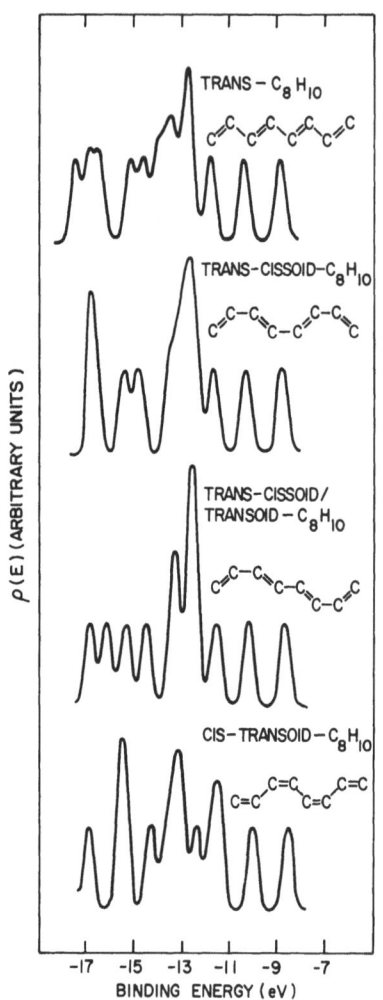

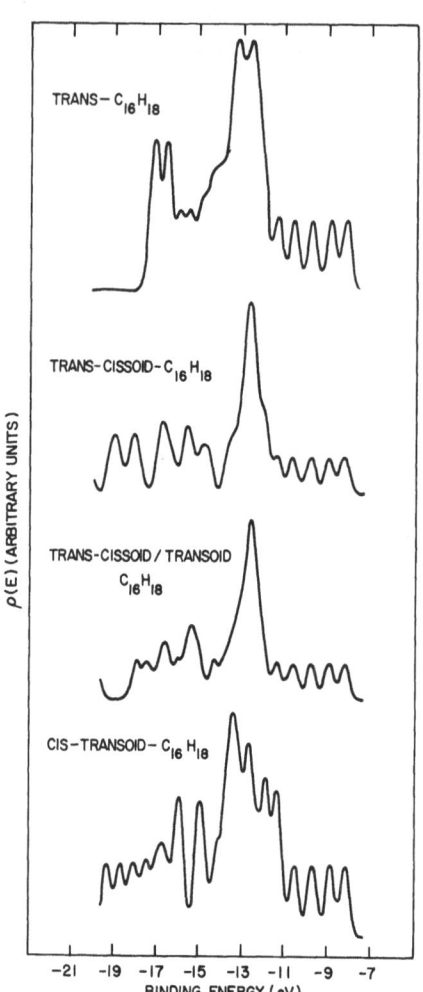

Fig.1: Calculated CNDO/S3 densities of valence electron states ("DOVS") for 1,3,5,7- octatetraene in its four possible planar sp^2 conformations. An intrinsic width of β = 0.3eV was utilized to obtain the DOVS from the calculated molecular orbital eigenvalue spectrum.

Fig.2: Calculated CNDO/S3 DOVS for 1,3,5,7,9,11,13,15-Hexadecaoctaene evaluated as described in the caption to Fig. 1.

are smaller than those between any of these conformations and the
cis -transoid conformation.

Another important aspect of photoemission from polyacetylene
is the assessment of the spatial extent of molecular cation states
generated by a photoionization event. For this purpose we show in
Fig. 3   the evolution   of the DOVS   of   trans - $H(CH)_n H$ as a
function of n. The   emergence   of   the π-electron   band in   the
energy region $-11eV \lesssim E \lesssim -7eV$ and of a σ-electron band in   the
region $-17eV \lesssim E \lesssim -11eV$ is evident from the figure.   For
typical solid-state width parameters, i.e., $\beta = 0.7eV$, the calculated DOVS remain invariant under increasing n for $n > 16$.   Thus, we
infer from the comparison of the calculated DOVS and the observed
photoemission spectrum shown in Fig. 3 that in the σ bands and the
center of the π band the molecular ion states extend over at
least 16-20 carbon atoms on the $(CH)_x$ backbone. The absence of the
predicted small shoulder at the low-binding-energy threshold of
the photoemission spectrum implies that the photoinduced cation
states at the upper edge of the $(CH)_x$ π-electron valence band are
localized by disorder (10). From the lower three panels of Fig. 3,
we estimate that the "band-edge" states are localized to within
$n \simeq 10$ carbon atoms, in contrast to the values $n \gtrsim 16$ characteristic of the center of the π and σ electron bands. Finally, comparing
the photoemission spectrum in Fig. 3 with the calculations for
$H(CH)_{16}H$ given in Fig. 2 indicates that the $(CH)_x$ chains in the
sample occur predominately the trans configuration over spatial
distances of the order of 20 carbon atoms: a result inferred
independently from the optical properties of the sample (16).   The
dependence on conformation of low-energy σ-electron molecular
cation states in oligomers also has been noted in other polymer
systems, especially polypropylene (21), by utilization of X-ray
valence electron photoemission spectroscopy which is much more
sensitive to these states than to the π-electron states.

Poly(1,6-heptadiyne)

The focus of our study of poly(1,6-heptadiyne) is the influence on the DOVS of the changes in conformation and electronic
structure wrought by the propyl side groups. We show in Fig. 4 the
evolution of the DOVS for poly(1,6-heptadiyne) as a function of n
in the series of model molecules trans - $H[CH(C_6H_7)]_n H$. Thus,
Fig. 4 for poly(1,6-heptadiyne) is the precise analog of Fig. 3
for polyacetylene. It is evident that the substituents induce
substantial changes in even the π-electron contributions to the
DOVS, leading to profound alterations in the DOVS relative to that
of   polyacetylene.   Consequently,   although   the   structure   of
the $(-CH=CH-)_m$ chain in the backbone of trans - poly(1,6-hepta-
diyne) is nearly identical in structure to that of polyacetylene,
its electronic structure is discernably different. We also pre-

Fig.3: Calculated CNDO/S3 DOVS
for a series of <u>trans</u> even
polyenes.   In the lower
four panels a value of
β = 0.3eV was used to ob-
tain the DOVS from the
CNDO/S3 eigenvalue spec-
trum, corresponding to
the analysis of photo-
emission from individual
molecules   in   the   gas
phase. The top panel con-
tains the measured ultra-
violet   photoemission
spectrum (16) of a film of
(CH)$_x$ shifted to higher
binding energies by 2.5eV
in order to preserve the
same energy scale in all
of the figures. Finally,
the panel below it contains the CNDO/S3 DOVS
for <u>trans</u>- H(CH)$_{16}$H evaluated   using   β =0.7eV
corresponding   to   the
solid-state.   If
β = 0.7eV, the DOVS for
H(CH)$_m$H   do   not   differ
noticeably from that for
H(CH)$_{16}$H for m > 16.

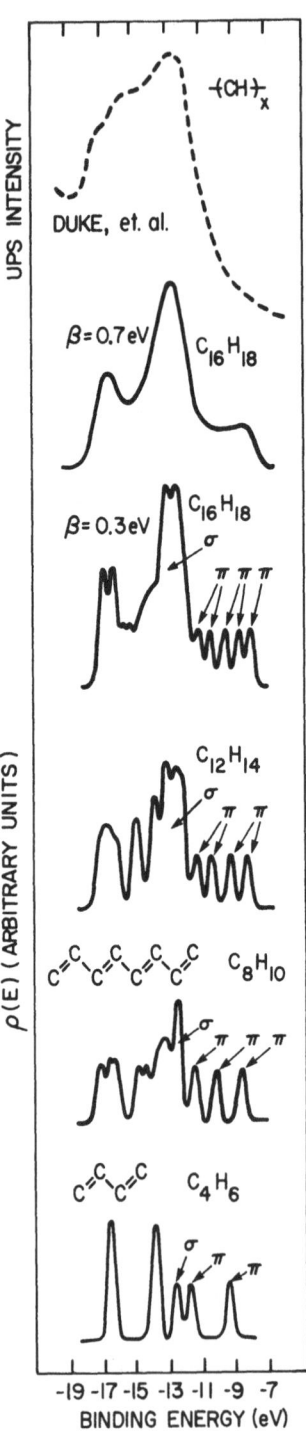

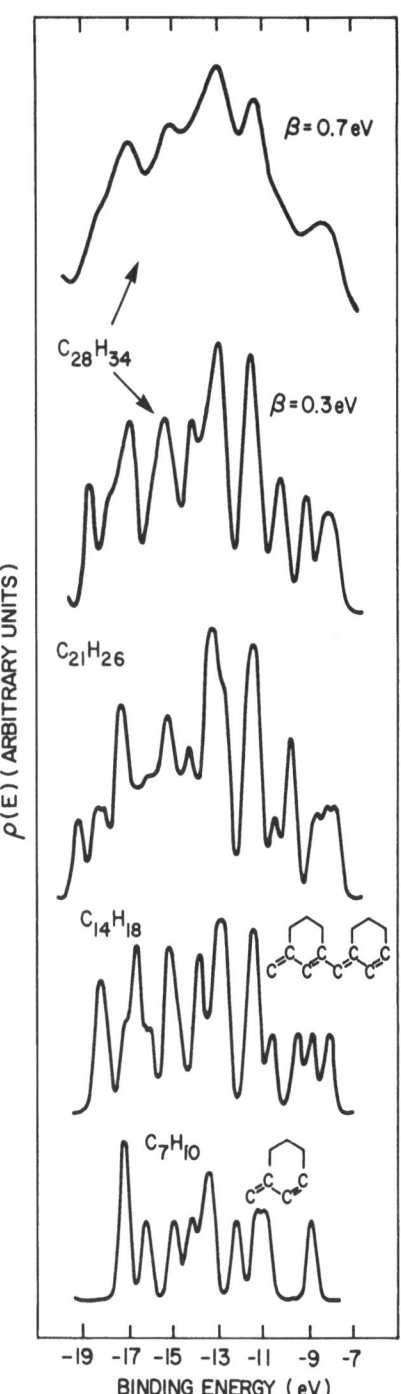

Fig.4: Calculated CNDO/S3 DOVS for a series of substituted even-polyenes, $H [CH(C_6H_7)]_n H$ (n=1,2,3 and 4) which become poly(1,6-heptadiyne) in the limit of large n. The value β =0.3eV for the width parameter was used to obtain the spectra shown in the lower four panels whereas the solid-state value, β =0.7eV, was utilized to construct the uppermost-panel for $H [CH(C_6H_7)]_4 H$.

dict a red shift by about 0.1 - 0.2 eV of the main $\pi \rightarrow \pi^*$ absorption band of poly(1,6-heptadiyne) relative to that of polyacetylene. Moreover, it is predicted to be more susceptible to chemical
attack because the backbone carbon species attached to side-group
carbons exhibit an electron defficiency of only 0.06e relative to a
value of 0.11 - 0.12e for the other two carbon species in the
backbone and to a uniform value of 0.09e for the carbons in $(CH)_x$.
The predicted increase in chemical instability of poly(1,6-hepta-
diyne) relative to polyacetylene is consistent with available data
(7), but the predicted changes in the DOVS and the red shift of the
lowest energy $\pi \rightarrow \pi^*$ transition have not yet been either verified
or disproved.

Poly(p-phenylene) and Related Polymers

The presence of the phenyl moieties in the backbone of poly-
(p-phenylene), designated by $(C_6H_4)_x$, poly(p-phenylene oxide),
designated by $(C_6H_4-O)_x$, and poly(p-phenylene sulfide), designated
by $(C_6H_4-S)_x$, results in π-electron molecular cation states
which are fundamentally different from those obtained for
polyacetylene chains. The doubly degenerate highest-energy
$e_{1g}(\pi)$ molecular orbital in benzene consists of a "non-bonding"
orbital (indicated by π_n) which exhibits nodes at carbon atoms
forming the links with the neighboring phenyl moieties, and a
"bonding" π orbital which exhibits maximum amplitude at these
linking carbon species. In diphenyl, for which the calculated DOVS
and observed gas-phase ultraviolet photoemission spectra (23)
are shown in Fig. 5, the non-bonding molecular ion states
remain near the energy (-9.25eV) of the corresponding molecular
ion state in benzene whereas the bonding orbitals form antibonding
(antisymmetric, π_a) and bonding (symmetric, π_b) molecular ion
states as noted in the figure. This behavior persists as longer
oligomers are built-up, leading to an intense narrow non-bonding
π band in the center of a "bonding" π band for $(C_6H_4)_x$. As shown
in the upper two panels of Fig. 5, the non-bonding π band is
predicted to be a prominent feature of the DOVS of $(C_6H_4)_x$,
although the bonding π band is so broad that its intensity gets
smeared out in the solid state. Neither the molecular-ion wave
functions nor their corresponding contributions to the DOVS are
directly analogous to the results for $(CH)_x$ for which the DOVS
corresponding to those shown in Fig. 5 were presented in Fig. 3.
Unfortunately no photoemission spectra are yet available to test
the validity of these predictions for $(C_6H_4)_x$.

The CNDO/S3 predictions for $(C_6H_4-O)_x$ and $(C_6H_4-S)_x$ are quite
comparable to those for $(C_6H_4)_x$ as evident from Figs. 6 and 7,
respectively. In these cases, however, the bonding π band is
split into two parts, depending upon whether the phenyl π-electron
linkages with the p orbitals on the chalcogens are bonding (π_b) or

Fig.5: Calculated CNDO/S3 DOVS for a series of polyphenyls, $H(C_6H_4)_n H$ which become poly(p-phenylene) in the limit of large n. The molecules are taken as planar, and comparison is made with available gas-phase ultraviolet photoemission spectra for benzene (22) and diphenyl (23). The gas-phase spectra in the figure are shifted to higher binding energy by about 1eV in order to retain the same energy scale in all the figures. Evaluation of the DOVS is described in the text and the caption for Fig.4.

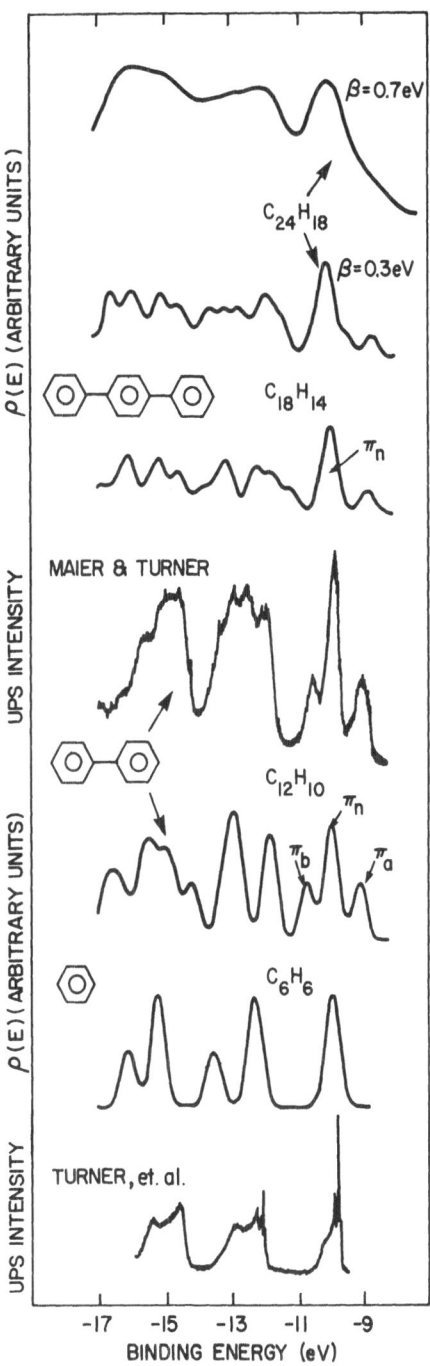

Fig.6: Calculated CNDO/S3 DOVS for the dimer and tetramer (n=2,4) derived from $HO(C_6H_4-O)_nH$ which corresponds to poly(p-phenylene oxide) in the limit of large n. A published structure of the polymer (24) was used to construct the oligomers. Evaluation of the DOVS is described in the text and the caption for Fig.4.

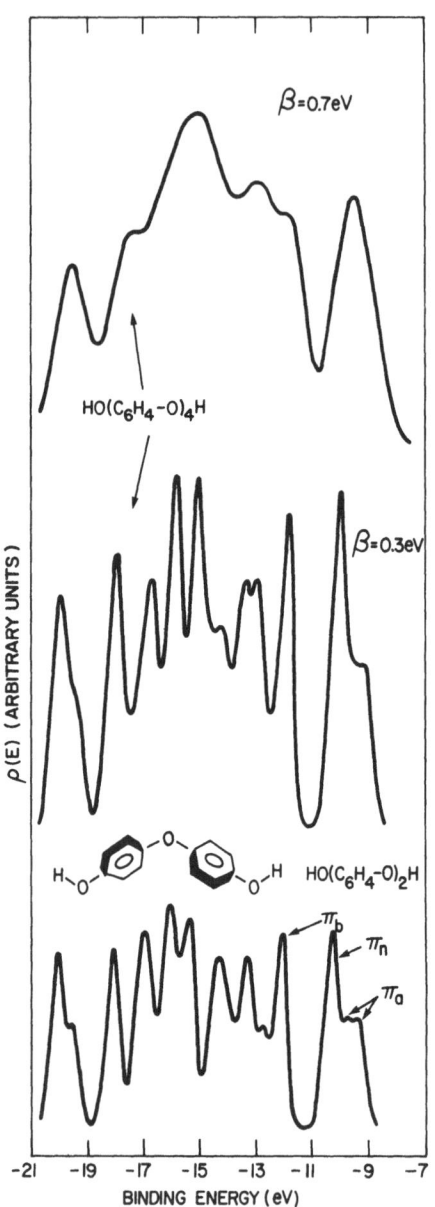

antibonding (π_a) in character. As in the case of $(C_6H_4)_x$, however, the most prominent π-electron contribution to the DOVS is predicted to be the strong, sharp peak caused by the non-bonding π molecular ion states. Also as in the case of $(C_6H_4)_x$, the lower-binding-energy antibonding π-electron states are smeared out to the point of being hard to identify in the DOVS in the solid state. For $(C_6H_4)_x$, $(C_6H_4-O)_x$ and $(C_6H_4-S)_x$ all three, however, the CNDO/S3 model predicts that the lowest-binding-energy molecular ion states are π-electron states which extend throughout the

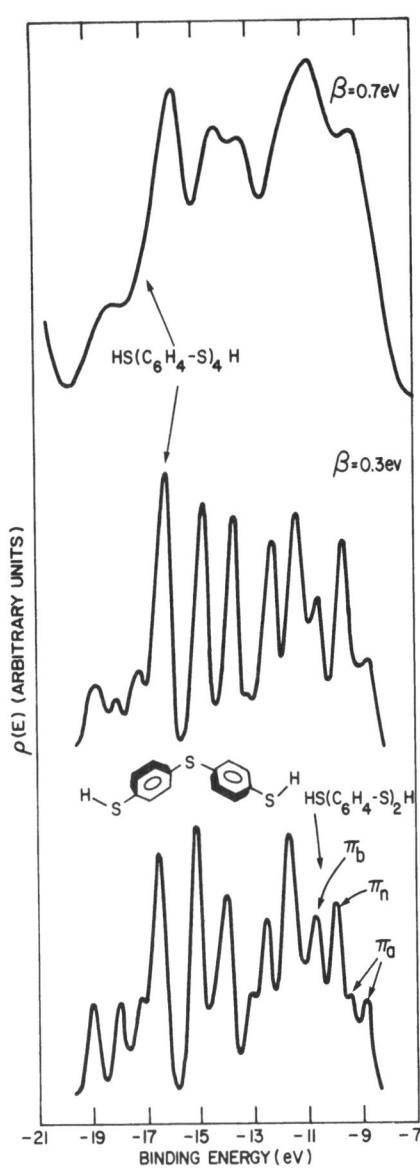

Fig.7: Calculated CNDO/S3 DOVS for the dimer and tetramer (n=2 and 4) derived from HO(C$_6$H$_4$-S)$_n$H which corresponds to poly(p-phenylene sulfide) in the limit of large n. A published structure of the polymer (25) was used to construct the oligomers. Evaluation of the DOVS is described in the text and the caption for Fig.4.

molecule. Therefore the doping of these materials with acceptors should produce electrical behavior comparable to that observed in polyacetylene. If the localized non-bonding π-electron states had exhibited the lowest binding energies, we would have anticipated difficulty in achieving high conductivities and mobilities via acceptor doping.

Poly(2,5-thienylene)

The contributions to the DOVS from the bonding π bands in ring

Fig.8: Calculated CNDO/S3 DOVS for a series of polythienyls, $H(C_4H_2S)_nH$ (n=1,2,3, and 4) which become poly(2, 5-thienylene) in the limit of large n. A symmetrized version of the structure of thiophene (26) and an antisymmetric planar orientation of the thienyl moieties (27) were utilized in the calculation. For thiophene, comparison with the ultraviolet photoemission spectrum of Baker et al. (28) is shown in the lower two panels. The photoemission spectrum is shifted to higher binding energies by 0.5eV in order to retain the same energy scale in all of the figures. Evaluation of the DOVS is described in the text and in the caption to Fig.4.

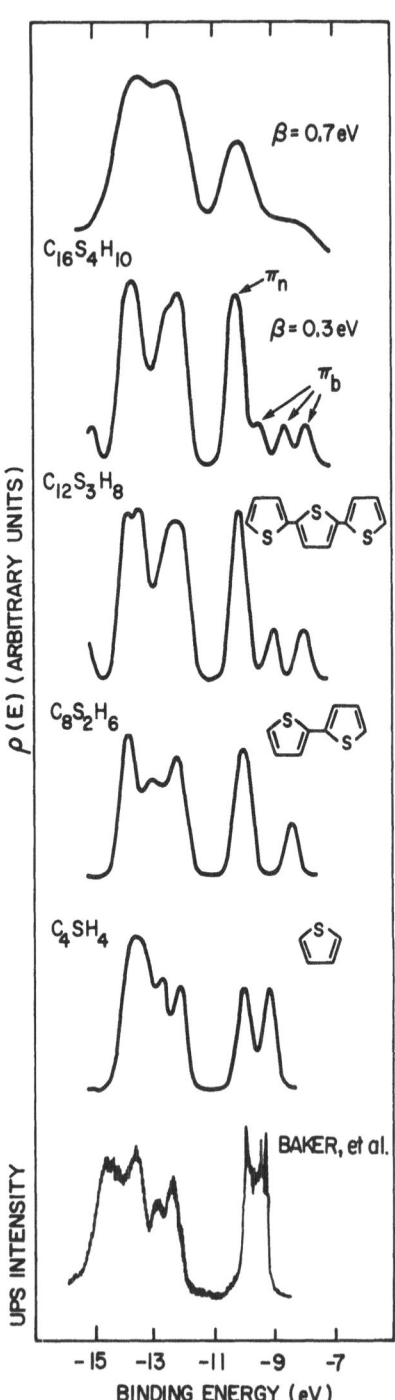

architectures like the polyphenyls can be rendered more similar to
the π bands in chain architectures like polyacetylene by utilizing asymmetric rings.   A good example of this possibility is
poly(2,5-thienylene) which is comprised of thiophene molecules in
a fashion analogous to the construction of poly(p-phenylene) from
benzene rings.   For thiophene the two highest-energy orbitals
(corresponding to the e_{1g} (π) orbitals in benzene) are π orbitals with a nodal plane through the sulfur and the center of the
back C-C bond, and with a nodal plane through the two carbon atoms
adjacent to the sulfur, respectively. These orbitals correspond
to the two isolated low-binding-energy ionizations between 9eV and
10eV evident in the lower two panels of Fig. 8.   The highest-
energy thiophene orbital generates a "bonding" π band in the
polythienyls nearly identical to the upper portion of the π band
in trans-cissoid/transoid polyenes, for which the corresponding
DOVS are shown in the second panel from the bottom in Figs. 1 and
2.   The lower portion of the polyene bonding π band is greatly
distorted in the polythienyls, however, because the totally symmetric π orbital in thiophene contains a substantial admixture
of the sulfur p_z orbital.

As in the case of the polyphenyls, the non-bonding π molecular ion state in thiophene (the ionization peak near -10eV binding
energy in Fig. 8) remains nearly invariant as polythienyls are
formed by joining multiple thiophene rings. In $(C_4SH_2)_x$ it becomes
a non-bonding π band ($π_n$) which is a prominent feature of the
calculated DOVS as shown in the upper four panels of Fig. 8.   In
this particular case, the non-bonding π band lies at the high-
binding-energy extremity of the low-binding-energy bonding π
band.   This behavior is similar to that observed at low binding
energies in the phenylene oxides and sulfides, as may be discerned
upon comparison of Fig. 8 with Figs. 6 and 7, respectively.

SYNOPSIS

Using the CNDO/S3 molecular orbital model (12), calculations
of the orbital eigenvalues and eigenfunctions have been performed
for oligomers of $(CH)_x$, $[CH(C_6H_7)]_x$, $(C_6H_4)_x$, $(C_4SH_2)_x$,
$[(C_6H_4)-O]_x$ and $[(C_6H_4)-S]_x$. In some cases optical absorption
bands have been calculated also via CI analyses based on these MO
eigenfunctions and eigenvalues. Our calculations permit the detailed interpretation of existing valence electron photoemission
data on these materials, the prediction of new photoemission and
optical absorption spectra, and the systematic identification and
classification of various influences of molecular architecture and
conformation on the energetics of   π electrons in linear-chain
polymers. Distinctions are drawn between localized and extended
π-electron   systems in isolated macromolecules which should aid
in the assessment of the nature of semiconductor-to-metal transitions in doped materials.   In particular, all of the polymers

considered herein exhibit extended π-electron states as the lowest-binding-energy molecular cation states induced by the photoionization of individual molecules, although evidence was presented that structural disorder localizes these states in polymer films.

REFERENCES

(1) C.K. Chiang, S.C. Gau, C.R. Fincher, Jr., Y.W. Park, A.G. MacDiarmid and A.J. Heeger, Appl. Phys. Lett. 33(1978) 18.

(2) C.K. Chiang, Y.W. Park, A.J. Heeger, H. Shirakawa, E.J. Louis and A.G. MacDiarmid, J. Chem. Phys. 69(1978) 5098.

(3) Y.W. Park, M.A. Druy, C.K. Chiang, A.G. MacDiarmid, A.J. Heeger, H. Shirakawa and S. Ikeda, J. Polym. Sci., Polym. Lett. 17(1979) 195.

(4) S.L. Hsu, A.J. Signorelli, G.P. Pez and R.H. Baughman, J. Chem. Phys. 69(1978) 106.

(5) D.M. Ivory, G.G. Miller, J.M. Sowa, L.W. Shacklette, R.R. Chance and R.M. Baughman, J. Chem. Phys. 71(1979) 1506.

(6) T. Yamamoto, K. Sanechika, and A. Yamamato, J. Polym. Sci. Polym. Lett. 18(1980) 9.

(7) H.W. Gibson, F.C. Bailey, J.M. Pochan, A.J. Epstein and H. Rommelmann, Organic Coatings and Plastics Chemistry Preprints 42(1980) 603.

(8) J.F. Rabolt, T.C. Clarke, K.K. Kanazawa, J.R. Reynolds, and G.B. Street, J. Chem. Soc. Chem. Commun.(1980) 347.

(9) R.R. Chance, L.W. Shacklette, G.G. Miller, D.M. Ivory, J.M. Sowa, R.L. Eisenbaumer and R.H. Baughman, J. Chem. Soc. Chem. Commun. (1980) 348.

(10) C.B. Duke, in "Extended Linear Chain Conductors," J.S. Miller, ed., (Plenum, New York, 1980), in press.

(11) C.R. Fincher, Jr., M. Ozaki, A.J. Heeger and A.G. MacDiarmid, Phys. Rev. B 19(1979) 4140.

(12) C.B. Duke, Int. Jour. Quant. Chem.: Quant. Chem. Symposm. 13(1979) 267.

(13) C.B. Duke, W.R. Salaneck, A. Paton, K.S. Liang, N.O. Lipari and R. Zallen, in "Structure and Excitations of Amorphous Solids," G. Lucovsky and F. Galeener, eds., (AIP, New York, 1976), pp. 23-30.

(14) C.B. Duke, Surface Sci. 70(1978) 674.

(15) K.L. Yip, N.O. Lipari, C.B. Duke, B.S. Hudson and J. Diamond, J. Chem. Phys. 64(1976) 4020.

(16) C.B. Duke, A. Paton, W.R. Salaneck, H.R. Thomas, E.W. Plummer, A.J. Heeger and A.G. MacDiarmid, Chem. Phys. Lett. 59(1978) 146.

(17) C.B. Duke, A. Paton, and T.J. Fabish, Chem. Phys. Lett. 49(1977) 133.

(18) C.B. Duke, W.R. Salaneck, T.J. Fabish, J.J. Ritsko, H.R. Thomas and A. Paton, Phys. Rev. B 18(1978) 5717.

(19) W.R. Salaneck, C.B. Duke, W. Eberhardt, E.W. Plummer, and H.J. Freund, Phys. Rev. Lett. 45(1980) 280.

(20) L.E. Sutton, ed., "Tables of Interatomic Distances and Configuration in Molecules and Ions" (Chemical Society, London, 1958).

(21) J.M. Andre, Polym. Preprints 21(1980) 127.

(22) D.W. Turner, C. Baker, A.D. Baker, and C.R. Brundle, "Molecular Photoelectron Spectroscopy" (Wiley Interscience, London, 1970), p. 271.

(23) J.P. Maier and D.W. Turner, Farad. Disc. Chem. Soc. 54(1972) 149.

(24) J. Boon and E.P. Magre, Makromol. Chem. 126(1969) 130.

(25) B.J. Tabor, E.P. Magre and J. Boon, Eur. Polym. Jour. 7(1971)1127.

(26) B. Bak, D. Christensen, L. Hansen-Nygaard, and J. Rastrup-Andersen, Jour. Mol. Spectros. 7(1961) 58.

(27) G.J. Visser, G.J. Heeres, J. Wolters and A. Vos, Acta. Cryst. B 24(1968) 467.

(28) A.D. Baker, D. Betteridge, N.R. Kemp, and R.E. Kirby, Anal. Chem. 42(1970) 1064.

THE EFFECT OF STRUCTURAL VARIABLES ON THE CONDUCTIVITY

OF IODINE DOPED POLYACETYLENE

Walter Deits, Peter Cukor and Michael Rubner

GTE Laboratories Incorporated
40 Sylvan Road
Waltham, MA 02154

INTRODUCTION

Many of the chemical and physical properties of polyacetylene have been shown to undergo dramatic changes after exposing the polymer to selected impurities or dopants. Many of these changes are thought to be brought about through the formation of charge-transfer complexes between polyacetylene and the doping species.[1] The properties of the polymer and of the charge-transfer complex have been investigated by infrared, visible, and ultraviolet optical spectroscopy,[2] as well as by photoelectron,[1,3] Raman,[1,3,4] x-ray diffraction,[1,3,5] and electron spin resonance[6-11] spectrographic techniques. Detailed studies on the morphology[12-14] and electrical properties[3,15,16] of the materials have also been undertaken. Inherent in all of these investigations has been the underlying importance of gaining a thorough knowledge of the effect of polyacetylene microstructure on the properties of the doped polymer.

Although many of the studies mentioned above have included excellent characterizations of the materials used, the majority have utilized polyacetylene samples prepared using techniques pioneered by Shirakawa et al.[12,13,17,18] in 1971. To date, however, there has been no comprehensive examination of structure/property relationships inherent to polyacetylene prepared using varied procedures. For this reason, and in order to gain a better understanding of the factors that influence the process of conduction in doped polyacetylene, we have undertaken a systematic study of polyacetylene prepared using a variety of methods under a number of different conditions.

Since the initial work done by Natta in 1958,[19] a great number of reports have been published concerning the synthesis and properties of polyacetylene. By changing catalysts, solvents, and reaction temperatures, Hatano[6] succeeded in preparing polyacetylene with varying degrees of crystallinity. In 1964, Tsuchida and coworkers[20] completely dehydrohalogenated polyvinyl chloride to form amorphous polyacetylene. Using a dilute catalyst system, Wnek and coworkers[16] developed a method for producing low density gel-like "foams" of polyacetylene similar in electrical properties to the Shirakawa material. Only recently, Voronkov et al.[21] reported on the utilization of molybdenum and tungsten halides and oxohalides as effective initiators for the polymerization of acetylene.

EXPERIMENTAL

Materials

All solvents were carefully dried and distilled under argon. Titanium (IV) butoxide (A), titanium (IV) isopropoxide (A), and titanium (IV) chloride (A) were vacuum distilled immediately prior to use. Tungsten (VI) chloride (A) and molybdenum (V) chloride (A) were used as received in sealed vials. Triethylaluminum (E) and ammonia (M) were used directly as received. Acetylene (M) was bubbled through a glass-packed sulfuric acid tower followed by passage through a KOH/3Å molecular sieve column and a cold trap (dry ice/acetone cooled) prior to use.

A: Alfa Division, Ventron Corporation; E: Ethyl Corporation; M: Matheson Corporation.

Polymerization Procedure

Three methods were used for preparing polyacetylene: Method A — A mixture of catalyst and solvent was exposed to gaseous acetylene (initial pressure ∿500 mm) for 72 hr according to the method of Wnek et al.[16] The resultant material was washed repeatedly with fresh solvent and dried at 10^{-3} mm for 72 hr.

Method B — Acetylene was bubbled for 2.5 hr through a mixture of catalyst and solvent. The resultant material was washed repeatedly with fresh solvent and dried at 10^{-3} mm for 72 hr.

Method C — Atactic polyvinyl chloride (inherent viscosity 1.02) was dehydrochlorinated in a mixture of sodamide, liquid ammonia, and tetrahydrofuran according to the method of Tsuchida and coworkers.[20] The resultant material was washed repeatedly with water and methanol and dried at 10^{-3} mm for 72 hr.

Doping Procedure

Samples were exposed to iodine vapor for 1.5 to 3.0 hr at a pressure of 0.1 mm. The maximum uptake of iodine was determined by weighing the samples at periodic intervals until a constant weight was obtained.

Measurements

Infrared spectra were recorded on a Perkin-Elmer 299B Infrared Spectrophotometer. X-ray diffraction patterns were recorded on a Phillips Vertical Diffractometer with a solid state scintillation detector. Scanning electron micrographs were obtained using a Jeolco U3 Instrument. Diffuse reflectance measurements were made on a Cary 17D Visible-Ultraviolet Spectrophotometer.

DC conductivities were measured on compressed pellets or films using standard four-point probe techniques.

RESULTS AND DISCUSSION

Changes in the morphology and microstructure of polyacetylene were brought about by varying the synthetic procedures used for the preparation of the polymers. The reaction conditions and synthetic techniques used are summarized in Table 1. Samples 1 through 5 were obtained as free-standing gels or films; samples 6 through 10 were isolated as powders and sample 11 was of a fibrous nature. Samples 1 through 10 could all be pressed into tough, shiny flakes or films. The polymers obtained were characterized by elemental analysis, infrared spectrophotometry, x-ray diffraction, scanning electron microscopy, diffuse reflectance in the visible-near infrared region, and dc conductivity after maximum iodine doping. All of the materials were found to contain less than 1.5% oxygen with the exception of sample 11 which contained up to 16% oxygen.

TABLE 1
PREPARATION OF POLYACETYLENE

Sample	Polymerization Method[a]	Catalyst	Catalyst Concentration (mole/l)	Solvent	Temperature (°C)
1[b]	A	$Ti(BuO)_4/4Et_3Al$	0.0125[e]	Toluene	−76
2	B	$Ti(BuO)_4/4Et_3Al$	0.0125[e]	Toluene	−76
3[c]	A	$Ti(BuO)_4/4Et_3Al$	0.0125[e]	Toluene	−76
4	A	$Ti(BuO)_4/4Et_3Al$	0.0125[e]	Toluene	20
5	A	$Ti(i—PrO)_4/$ $4Et_3Al$	0.0125[e]	Heptane	−76
6	B	$WCl_6/\frac{1}{2}H_2O$	0.004[f]	Toluene	7
7	B	$MoCl_5/\frac{1}{2}H_2O$	0.006[g]	Toluene	−76
8	B	$TiCl_4/1.2Et_3Al$	0.0073[e]	Toluene	−76
9	B	$TiCl_4/1.2Et_3Al$	0.0073[e]	Toluene	20
10	B	$TiCl_4/1.2Et_3Al$	0.0073[e]	Heptane	20
11	C	$NaNH_2$[d]	0.84	THF/NH_3	50

(a)   As designated in Experimental Section.
(b)   Essentially identical to the material studied by Wnek et al.[16]
(c)   Sample 1 annealed at 198 °C for 1.5 hr.
(d)   Dehydrohalogenating reagent.
(e)   Based on Ti.
(f)   Based on W.
(g)   Based on Mo.

Table 2 summarizes some of the more significant properties of the various polyacetylene samples. It is evident that the conductivity of iodine-doped polyacetylene is determined to an appreciable extent by the morphology and microstructure of the starting material. As reported by Shirakawa et al.[22] and others,[2] the geometrical configuration of the polymer backbones seems to play a moderate role in determining the conductivity, with the most conducting materials also containing the greatest amount of the cis isomer. Materials with an essentially 100% cis configuration have conductivities ranging from a low of 150 to a high of 700 $\Omega^{-1}cm^{-1}$. The introduction of a moderate amount of the trans isomer lowers the conductivity by a factor of three or more to 50 to 80 $\Omega^{-1}cm^{-1}$. Further increases in the amount of the trans isomer do not have a significant effect as was shown by converting a predominantly cis material (sample 1) to one of 100% trans content (sample 3) by thermal isomerization according to the procedure of Ito et al.[13]

TABLE 2
PROPERTIES OF POLYACETYLENE

Sample[a]	Cis Content (%)[b]	X-ray Peak Width[c]	Morphology[d]	Conductivity of Iodine Doped Material $(\Omega^{-1}cm^{-1})$[e]
1	74	1.2	Fibrillar	105 — 250
2	98	1.0	Fibrillar	500 — 700
3	0	1.8	Fibrillar	50 — 80
4	28	1.1	Fibrillar	50 — 80
5	29	1.8	Fibrillar	50 — 80
6	24	5.0	Globular w/Short Fibrils	50 — 80
7	24	2.0	Globular	70 — 100
8	3	5.0[f]	Globular w/Short Fibrils	6×10^{-3}
9	7	3.5[f]	Globular	3×10^{-4}
10	0	3.5[f]	Globular	5×10^{-4}
11	0	3.5[f]	Fibrillar	6×10^{-5}

(a) As designated in Table 1.
(b) Calculated by method established in Ref. 12.
(c) Relative peak width at one-half peak height at $2\theta = 23°$ to $25°$ in x-ray diffraction pattern.
(d) As revealed by scanning electron microscopy.
(e) Measured at maximum iodine uptake (iodine content = 15 to 25 mole %).
(f) A second broad peak present at $2\theta = 16°$.

Although different morphologies have been observed (Figure 1), the conductivity of iodine-doping polyacetylene does not seem to be influenced to a great extent by the gross morphology of the material as determined by scanning electron microscopy. The materials with the greatest conductivity do seem to have the

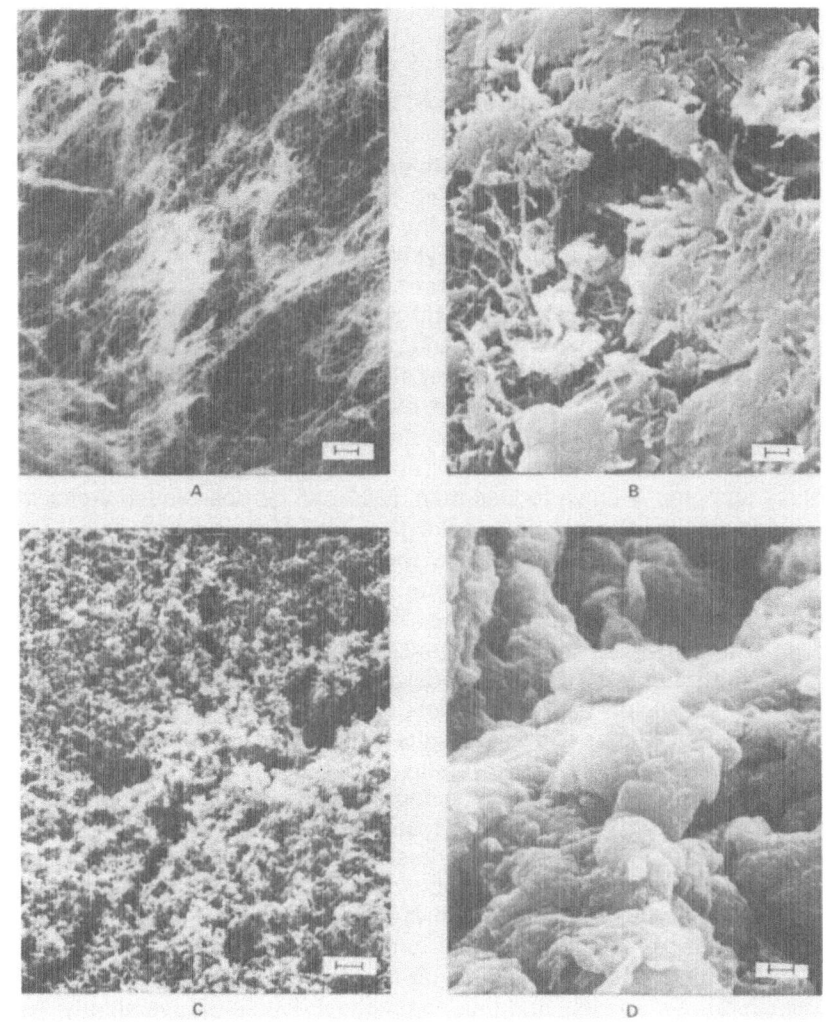

Figure 1. Scanning electron micrographs of polyacetylene. A: sample 2; B: sample 5; C: sample 6; D: sample 7. Scale bar is 2 μm.

most fibrillar morphology, however, with the conductivity increasing as the diameter of the fibrils decreases. The finest fibrils observed have been on the order of 200Å to 300Å in diameter. Those materials exhibiting the lowest conductivity upon iodine-doped showed an essentially "globular" morphology with a complete absence of fibrils. An exception to this trend can be seen in sample 7, which exhibited a predominantly "globular" morphology while remaining highly conducting. The reasons for these differences are not apparent at this time; however, those samples exhibiting a predominantly globular morphology were isolated as fine powders in contrast to those materials exhibiting a fibrillar morphology which were obtained as gels or films. It is quite possible that the interior of the powder particles are fibrillar in nature as evidenced by the short fibrils observed in sample 8; however, this has not yet been conclusively demonstrated. In addition, sample 11, polyacetylene prepared by dehydrohalogenating polyvinyl chloride, exhibited a conductivity of 6×10^{-5} $\Omega^{-1}cm^{-1}$ after doping with iodine while having a predominantly fibrillar morphology with the fibrils nearly 1 μm in diameter. In this case, however, the nature of the polyvinyl chloride starting material may have influenced the morphology of the resultant sample in some manner. It should be noted that although a number of different morphologies have been attributed to different polyacetylene samples prepared in various investigations, we have found that significantly different morphological characteristics can be observed in the same sample depending on the area examined. Figure 2, for example, shows typical micrographs of polyacetylene prepared by bubbling acetylene gas through a homogeneous catalyst solution of $Ti(BuO)_4/AlEt_3$ in toluene. An essentially featureless topography down to less than 500Å can be observed in areas where the polymerization takes place next to the glass wall of the reactor vessel. The exterior surface of the as-formed gel, on the other hand, shows a rough three-dimensional nearly globular appearance with some evidence of a fibrillar character. Upon examining the interior of the gel, however, one finds a material composed entirely of fine fibrils 200Å to 300Å in diameter. This phenomenon is similar to that attributed by Karasz and coworkers[14] to factors such as solvent effects, thermal expansion and contraction, local heating, and fusion during the polymerization of acetylene and to solubility differences during the washing of poly(1,6-heptadiyne) by Gibson and coworkers.[23]   The conductivity of iodine-doped polyacetylene, although related to some degree to both the geometrical configuration and the morphology of the starting material, seems to be much more closely related to the degree of crystallinity present in the undoped material. Relative degrees of crystallinity were estimated by measuring the peak widths at one-half the peak heights at $2\theta = 23°$ to $25°$ in the x-ray diffraction patterns of the different polyacetylene samples. Typical diffraction patterns are illustrated in Figure 3. The most conductive sample (2) was found to have the narrowest and most intense diffraction peak and thus the highest degree of crystallinity. As the peak intensities decreased and the peak widths increased, the conductivity dropped significantly. In addition to a weak broad diffraction peak at $2\theta = 23°$ to $25°$ indicative of decreased crystallinity, samples 8, 9, 16 and 11 also had a broad peak centered at $2\theta = 16°$ that was not observed in any other samples. A broad peak in this region has been reported to be due to scattering from amorphous regions of polyacetylene[20,24] as well as other polymers.[25] The presence of amorphous scattering in a number of the samples implies that there are distinct differences in the fine structures of the various polyacetylene preparations. Polyethylene, and other

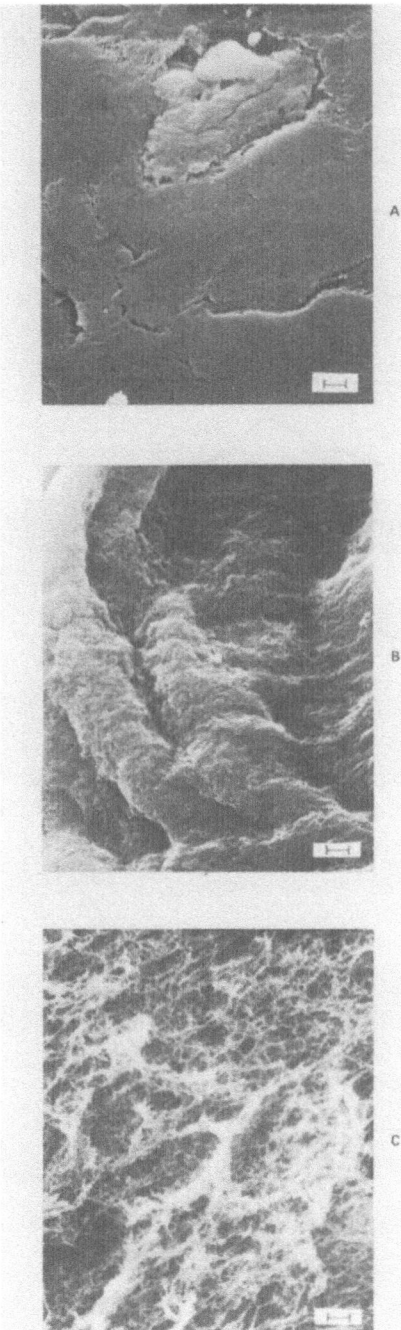

Figure 2. Scanning electron micrographs of polyacetylene (sample 2). A: surface of gel adjacent to glass reactor wall; B: surface of gel away from wall; C: interior of gel. Scale bar is 2 μm.

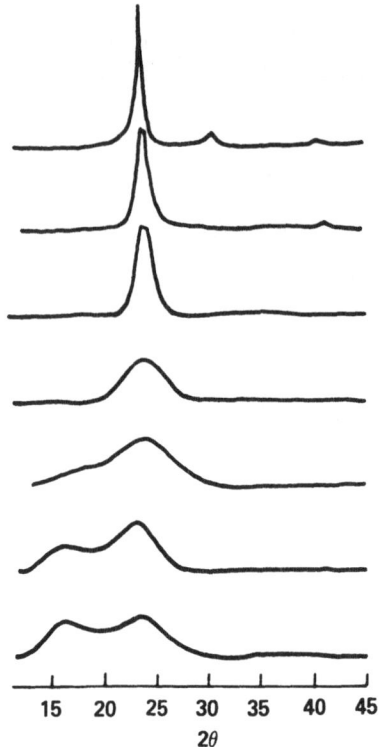

Figure 3. X-ray diffraction patterns of polyacetylene. A: sample 2; B: sample 5; C: sample 7; D: sample 6; E: sample 8; F: sample 9; G: sample 11.

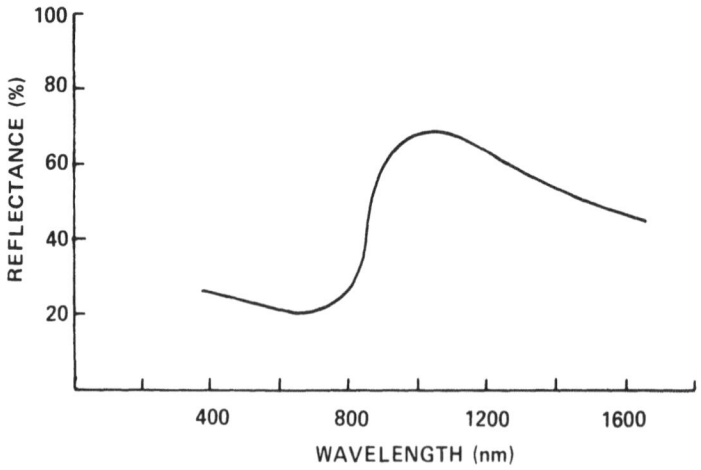

Figure 4. Diffuse reflectance spectra of polyacetylene.

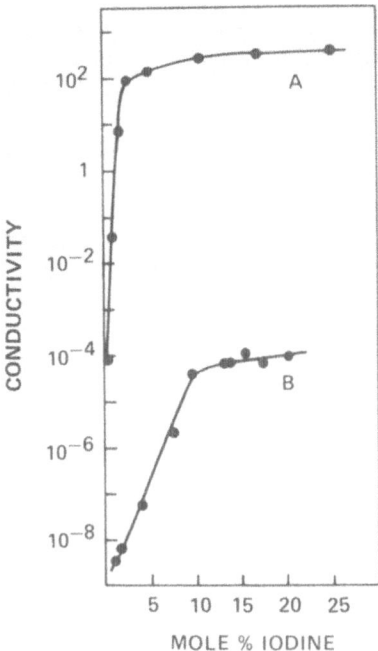

Figure 5. Conductivity of polyacetylene as a function of iodine concentration. A: sample 1; B: sample 11.

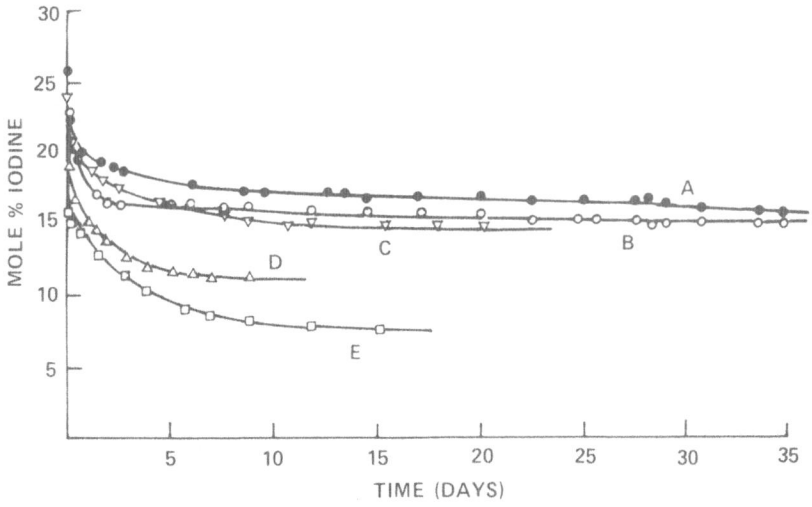

Figure 6. Iodine content of polyacetylene exposed to dynamic vacuum. A: sample 1; B: sample 5; C: sample 9; D: sample 8; E: sample 11.

semicrystalline polymers displaying shoulders or broad peaks in their x-ray diffraction patterns at $2\theta = 16°$ to 19°, for example, have been shown[25] to crystallize in lamallae with alternating amorphous layers to form superlattice structures with high degrees of long-range order. Polymers with rigid backbones, such as Kevlar, however, show only single intense diffraction peaks at $2\theta = 21°$ corresponding to crystalline regions oriented in an extended-chain rather than a lamellar form.[26,27]

In our investigation, we found that the appearance of the broad peak at $2\theta = 16°$ coincided with greatly reduced conductivities with the magnitude of the conductivity inversely proportional to the intensity. This suggests that conductance in doped polyacetylene is greatly influenced by the degree of long-range order existing between the polymer chains. These conclusions are further substantiated by recent x-ray diffraction studies done by Akaishi and coworkers[5] who found degrees of crystallinity up to 0.81 in polyacetylene samples prepared using a $Ti(BuO)_4/Et_3Al$ catalyst solution. The polyacetylene crystalline regions were proposed as being more disordered than typical polyethylene crystals owing to the higher rigidity of the polyacetylene chains. Polyacetylene was also shown to exist with an extended-chain type crystal morphology. It was also suggested that the high electrical conductivity observed in bulk polyacetylene is undoubtedly facilitated by both the high degree of crystallinity and the extended-chain crystal morphology. Although Akaishi and coworkers did not observe a second peak in the x-ray diffraction pattern at $2\theta = 16°$, the appearance of this peak in some of our preparations with corresponding reduced conductivities gives further evidence for the importance of long-range order in polyacetylene conduction.

Diffuse reflectance studies in the visible-near infrared region of the spectrum showed only a single broad absorption at wavelengths less than 980 nm (Figure 4) in agreement with other studies.[2] No discernible differences were found between the samples indicating that no major differences in chain length or degree of crosslinking were present in the various materials studied.

Although the conductivities of the iodine-doped polyacetylene samples examined in this study have been shown to vary over several orders of magnitude, all of the materials show a similar dependence on dopant concentration including the presence of a sharp metal to insulator transition as is illustrated in Figure 5 for samples 1 and 11. Equivalent behavior upon exposure to dynamic vacuum is also observed as is shown in Figure 6. These similarities point to the conclusion that the differences in conductivities observed were not due to inherent differences in the doping process itself, but were rather due to differences in the mobility of the charges resulting from microstructural and morphological changes brought about by the various preparative techniques used.

CONCLUSIONS

The conductivity of iodine-doped polyacetylene has been found to be related to the morphology, crystallinity, and microstructure of the polymer. The closest relationship is with the crystallinity, with the highest conductivity associated with the most crystalline materials. Conversely, low levels of conductivity coincide with the appearance of a broad peak at $2\theta = 16°$ in the x-ray diffraction pattern which is attributed to amorphous regions in the polyacetylene samples.

ACKNOWLEDGMENTS

The authors would like to acknowledge the invaluable assistance of Harriet Jopson in performing many of the experiments in this study and of the Materials Evaluation Group of GTE Labs for much of the analytical data.

REFERENCES

1. S.L. Hsu, A.J. Signorelli, G.P. Pez and R.H. Baughman, *J. Chem. Phys. 69*, 106 (1978).

2. C.K. Chiang, A.J. Heeger and A.G. MacDiarmid, *Ber. Bunsenges. Phys. Chem. 83*, 407 (1979).

3. R.H. Baughman, S.L. Hsu, G.P. Pez and A.J. Signorelli, *J. Chem. Phys. 68*, 5405 (1978).

4. H. Temkin, S. Lefrant, J. Burlich and D.B. Fitchen, *Bull. Am. Phys. Soc. 23*, 306 (1978).

5. T. Akaishi, K. Miyasaka, K. Ishikawa, H. Shirakawa and S. Ikeda, *J. Polym. Sci. Polym. Phys. 18*, 745 (1980).

6. M. Hatano, S. Kambara and S. Okamoto, *J. Polym. Sci. 51*, 526 (1961).

7. H. Shirakawa, T. Ito and S. Ikeda, *Makromol. Chem. 179*, 1565 (1978).

8. I.B. Goldberg, H.R. Crowe, P.R. Newman, A.J. Heeger and A.G. MacDiarmid, *J. Chem. Phys. 70*, 1132 (1979).

9. A. Snow, P. Brant, D. Weber and N.L. Yang, *J. Polym. Sci. Polym. Lett. 17*, 263 (1979).

10. J.C.W. Chien, F.E. Karasz, G.E. Wnek, A.G. MacDiarmid and A.J. Heeger, *J. Polym. Sci. Polym. Lett. 18*, 45 (1980).

11. M. Schwoerer, U. Lauterbach, W. Muller and G. Wegner, *Chem. Phys. Lett. 69*, 359 (1980).

12. T. Ito, H. Shirakawa and S. Ikeda, *J. Polym. Sci. Polym. Chem. 12*, 11 (1974).

13. T. Ito, H. Shirakawa and S. Ikeda, *J. Polym. Sci. Polym. Chem. 13*, 1943 (1975).

14. F.E. Karasz, J.C.W. Chien, R. Galkiewicz, G.E. Wnek, A.J. Heeger and A.G. MacDiarmid, *Nature 282*, 286 (1979).

15. D.J. Berets and D.S. Smith, *Trans. Faraday Soc. 64*, 823 (1968).

16. G.E. Wnek, J.C.W. Chien, F.E. Karasz, M.A. Druy, Y.W. Park, A.G. MacDiarmid and A.J. Heeger, *J. Polym. Sci. Polym. Lett. 17*, 779 (1979).

17. H. Shirakawa and S. Ikeda, *Polym. J. 2*, 231 (1971).

18. H. Shirakawa, T. Ito and S. Ikeda, *Polym. J. 4*, 460 (1973).
19. G. Natta, G. Mazzanti and P. Corradini, *Atti. Acad. Nazl. Linci Rend. Classe Sci. Fis. Mat. Nat. 25*, 3 (1958).

20. E. Tsuchida, C. Shih, I. Shinohara and S. Kambara, *J. Polym. Sci. A2*, 3347 (1964).

21. M. Voronkov, V. Pukhnarevich, S. Suchchinskaya, V.Z. Annenkova, V.M. Annenkova and N. Andreeva, *J. Polym. Sci. Polym. Chem. 18*, 53 (1980).

22. H. Shirakawa, T. Ito and S. Ikeda, *Macromol. Chem. 179*, 1565 (1978).

23   H. Gibson, F. Bailey, J. Pochan, A. Epstein and H. Rommelmann, *Org. Coat. Plast. Chem. ACS 42*, 603 (1980).

24. M. Hatano, *Zasshi Kogyo Kagaku 65*, 723 (1963).

25. S. Krimm and A.V. Tobolsky, *J. Polym. Sci. 7*, 57 (1951).

26. K. Yabuki, H. Ito and T. Oota, *Senigakkaishi 31*, T-524 (1975).

27. K. Yabuki, H. Ito and T. Oota, *Senigakkaishi 32*, T-55 (1976).

STUDIES IN CONDUCTING POLYMERS

G.W. Wnek, J. Capistran, J.C.W. Chien, L.C. Dickinson,
R. Gable, R. Gooding, K. Gourley, F.E. Karasz,
C.P. Lillya, and K.-D. Yao

Departments of Polymer Science and Engineering, and
Chemistry, Materials Research Laboratory, University
of Massachusetts, Amherst, Massachusetts 01003

SECTION I: MOLECULAR WEIGHTS OF POLYACETYLENE

Introduction

Characterization of polyacetylene in terms of the classical polymer parameters - molecular weight, molecular weight distribution, branching, crosslinking, structural and isomeric irregularities, end groups, etc. - is still almost totally lacking. Nor is it known how these parameters vary with time during the polymerization process or by post-synthesis reaction with further adventitious or deliberately introduced reactants. It is clear that these factors must also effect the crystallinity and morphology of the resultant polyacetylene and hence ultimately the electrical properties.

The solvent intractability of $(CH)_x$ provides the most serious obstacle to the characterization of the polymer. To overcome this problem, methods have been described in which $(CH)_x$ is hydrogenated to yield, in principle, characterizable polyethylene[1,2]. Analogous reaction schemes, e.g. hydrochlorination, could be envisaged for the same purpose. These techniques obviously rest on the unvalidated premise that the resulting polymer accurately reflects the initial material in every respect save the reacted double bond.

An alternative approach is the radioactive labelling of the polymer chains by appropriate quenching reactions. This technique has been widely used to characterize Ziegler-Natta catalyst mechanisms[3,4,5]. The radioactive quenching reagents used in the

present study were carbon monoxide (^{14}CO) and methanol (CH_3O^3H).
The former reagent inserts into an active chain as follows:

$$- \overset{|}{\underset{|}{Ti}} - P\text{\textasciitilde\textasciitilde\textasciitilde} + \text{*CO} \rightarrow - \overset{|}{\underset{|}{Ti}} - \overset{O}{\overset{\|}{\text{*C}}} - P\text{\textasciitilde\textasciitilde\textasciitilde}$$

The chain, end-capped with the "hot" CO may then be cleaved with
"cold" methanol:

$$- \overset{|}{\underset{|}{Ti}} - \overset{O}{\overset{\|}{\text{*C}}} - P\text{\textasciitilde\textasciitilde\textasciitilde} + CH_3OH \rightarrow - \overset{|}{\underset{|}{Ti}} - 0 - CH_3 + H - \overset{|}{\underset{\underset{O}{\|}}{C^*}} - P\text{\textasciitilde\textasciitilde\textasciitilde}$$

In contrast to *CO, tritiated methanol will react with chains
growing on Al in addition to Ti sites:

$$- \overset{|}{\underset{|}{Ti}} - P\text{\textasciitilde\textasciitilde\textasciitilde} + CH_3O\text{*H} \rightarrow - \overset{|}{\underset{|}{Ti}} - OCH_3 + \text{*H} - P\text{\textasciitilde\textasciitilde\textasciitilde}$$

and

$$\overset{|}{\underset{|}{Al}} - P\text{\textasciitilde\textasciitilde\textasciitilde} + CH_3O\text{*H} \rightarrow \overset{|}{\underset{|}{Al}} - OCH_3 + \text{*H} - P\text{\textasciitilde\textasciitilde\textasciitilde}$$

Thus, as a first approximation, we may expect CO quenching
to yield an upper limit to the molecular weight; the ratio of molecular weights found in the *CO and CH_3O*H reactions will also
provide information about the transfer reactions.

Experimental

Polymerizations were carried out in 200 cm^3 glass pressure
reactor bottles sealed with crown caps and live-rubber septa.
Generally four parallel reactions were run with the bottles
attached via 22-gauge needles to a high vacuum manifold for overnight evacuation. Vessels were vigorously stirred during polymerization by magnetic bars. Catalyst stock solutions were prepared as 0.25 M Ti(OBu)$_4$ with a 4:1 AlEt$_3$:Ti ratio. This concentration is two orders of magnitude lower than that commonly employed to produce C$_2$H$_2$ film. Scrupulously washed and dried toluene stored with continuous reflux over CaH$_2$ was used. Schlenk
techniques and gas tight syringes were used to limit the possibility of O$_2$ and H$_2$O interference. Each polymerization bottle contained 25 μmole of titanium catalyst in 20 ml of toluene. Acetone
and H$_2$O were removed from the acetylene by passing the gas
through two scrubbers filled with H$_2$SO$_4$ and a tower of P$_2$O$_5$.
Acetylene in a 2 ℓ ballast bulb was deoxygenated by two freeze-
pump-thaw cycles. Acetylene pressures were never more than 0.9
atmosphere in order to limit the explosion hazard. Acetylene was
admitted to the reaction vessels via the scrub line in order to
maintain that pressure throughout the reaction time. Polymeriza-

tions were allowed to proceed for an hour and a half at ambient temperature before quench unless otherwise specified.

^{14}CO was purchased as 1 millicurie (ICN) and diluted into one liter of purified CO (Matheson Co.). A modified Toeppler pump permitted small increments of gas to be admitted into a chamber accessible through a rubber septum by gas tight syringe. CH_3OT labelled methanol was prepared by diluting 10 μl of H_2O (5 curies/ml from New England Nuclear Corp.) into 100 ml of absolute methanol. The methanol was dried by reflux and distillation from magnesium methoxide. CH_3O*H was stored in a crown capped, live rubber septum sealed bottle fitted with a Schlenk side arm for passing argon through during access.

Analysis of the ^{14}C and ^{3}H contents of the polymers was performed by the assay service of New England Nuclear Corp., Boston, Mass., by combustion of the $(CH)_x$ to $*CO_2$ or $*H_2O$ and counting the absorbed species by standard scintillation techniques. Background was typically 1-3% of the total counts. Ashed weight of samples was typically 1% of the total weight (Galbraith Laboratories, Knoxville, Tenn.). After radio-labelling polymer was extracted with three washes of 50 ml each of 10% conc. HCl in unlabelled methanol followed by three washes of absolute methanol and dried in vacuo overnight.

Results and Discussion

The experiments described were carried out in the time regime in which the net yield was continuing to increase linearly, Fig. 1. This yield contains a mixed population of chains attached to either Al or Ti sites as well as chains removed from the catalyst by disproportionation or other reactions.

The CH_3O*H reactions were carried out by pumping off the reacting acetylene and then quenching by 0.2 ml of radioactive methanol. The time of exposure to the CH_3O*H varied from 0.2 to 36 hrs. The activity of the extracted polymer, measured as specific disintegrations per unit time, was 1085 ± 100 dpm/mg, Fig. 2. This result yields a molecular weight of 41,200. However, this value must be corrected for the anticipated kinetic isotope effect. We have evaluated this by varying the amount of labelled CH_3O*H added. Preliminary results from this experiment, indicate a correction factor of 1.9 which lowers the calculated molecular weight to 22,000.

These results were consistent with the quenching experiments carried out using *CO. In the studies in which the C_2H_2 was removed before labelled quenching reagent was added an initial increase was followed by a plateau at about 200 dpm/mg for exposures to *CO longer than 3 hours. In a second series in which the

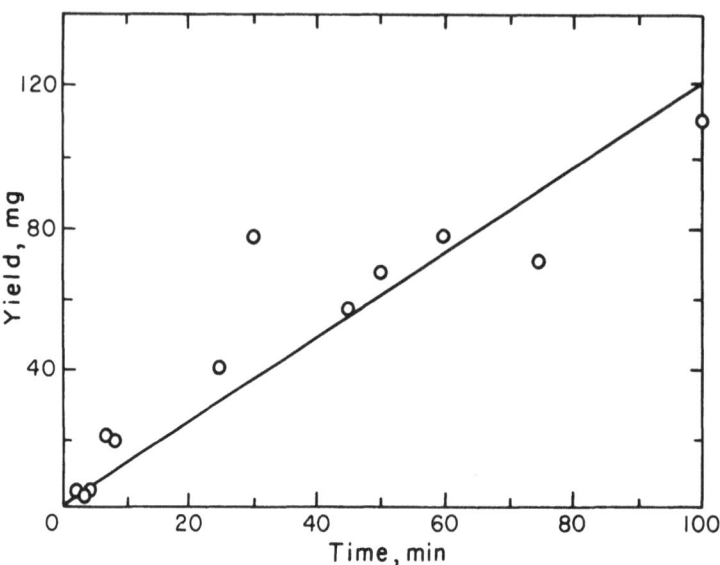

Figure 1: Yield of polyacetylene as function of polymerization
           time.

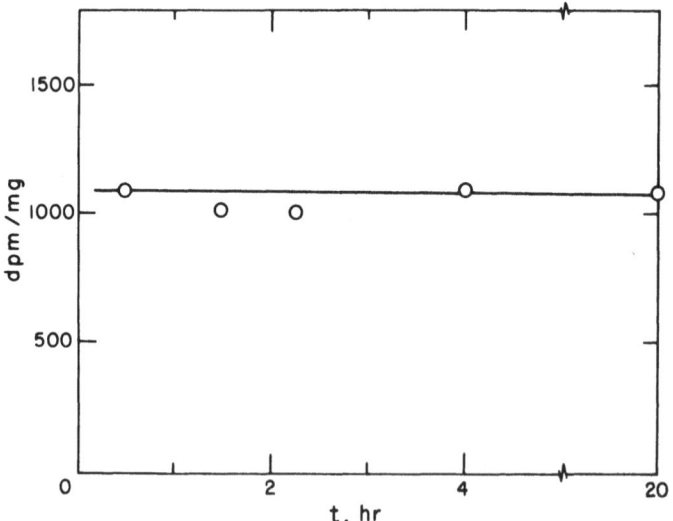

Figure 2: Specific activity, dpm/mg, of tritiated polymer as
           function of exposure time to tritiated methanol.

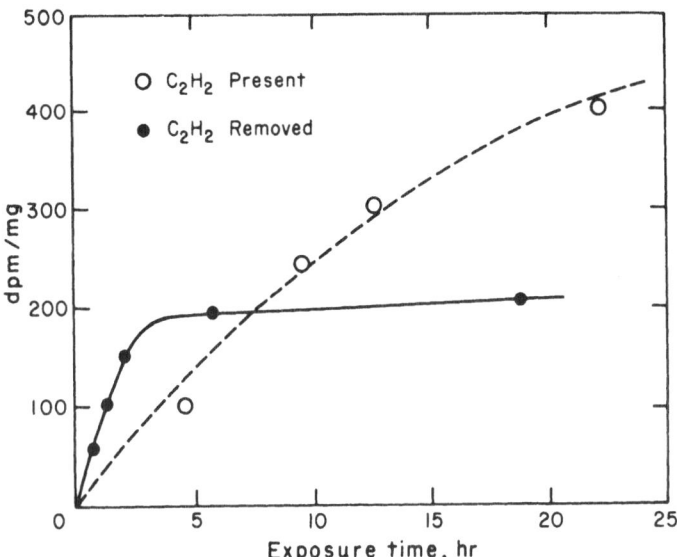

Figure 3: Specific activity of *CO quenched polymer as function
           of exposure time.

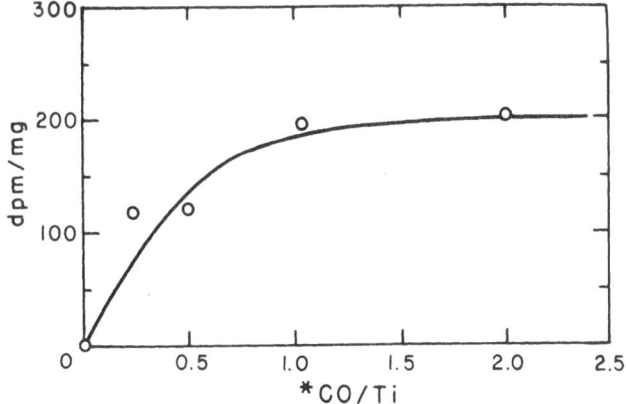

Figure 4: Specific activity of *CO quenched polymer as function
           of *CO/Ti.

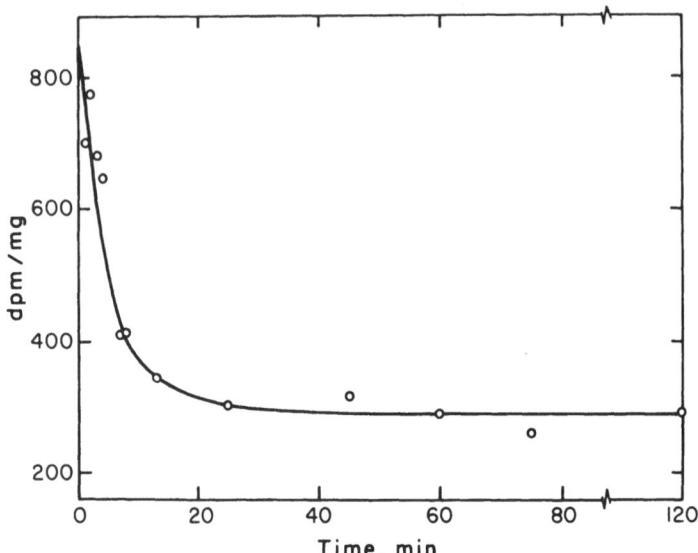

Figure 5: Specific activity of *CO quenched polymer as function
of time of polymerization before addition of *CO.

C_2H_2 was not removed a further increase in activity could be observed implying an additional statistical incorporation of *CO in
the chains growing from Ti sites, Fig. 3.

The amount of *CO required to produce this asymptotic activity in the resulting polymer was also studied. Fig. 4 shows that
this point was reached with a *CO/Ti ratio of greater than 1.
Further information was obtained by varying the polymerization
time at which the *CO was added to the reaction mixture. Fig. 5
indicates that transfer reactions significantly lower the activity
in the resulting polymer from the initial reaction rates. If the
activity is extrapolated to zero time (~800 dpm/mg) an upper limit
of molecular weight of approximately 55,000 is found. By comparing this to the value obtained from the labelled methanol quench
experiments we obtain a chain transfer ratio of 2.5. The propagation rate constant, k_p, can also be evaluated from the data with
a result of approximately $k_p \sim 4.3$ 1 mol^{-1} sec^{-1}.

We conclude from these somewhat preliminary results that the
radioactive labelling technique can provide consistent data for
the molecular weight of $(CH)_x$ under strictly controlled conditions.
These do not necessarily correspond to those used in the now standard preparation of fibrillar films and/or gels. The method is
not without its problems or ambiguities but also offers great versatility in terms of quenching agents. By varying these and
polymerization conditions systematically it will be possible to

obtain reliable molecular weight data as well as information about polymerization mechanisms.

SECTION II: ACETYLENE-METHYLACETYLENE COPOLYMERS

Introduction

The recent demonstration[6],[7] of the attainment of metal-like properties of polyacetylene upon chemical doping has stimulated a fundamental interest in this material. Polyacetylene, as well as other conducting polymers,[8],[9],[10] may be anticipated to have tremendous technological potential in view of the light weight advantages and low production costs of polymers in general as compared to other materials. However, as has been noted above, the intractability of most conducting polymer systems precludes rigorous characterization at the molecular level and attempts at processing by conventional methods.

The simplest approach toward obtaining a "modified" $(CH)_x$ backbone is the polymerization of substituted acetylenes (e.g., methyl-[11] and phenylacetylene[12]). The resulting polymers are attractive in view of their solubility in common organic solvents. However, upon doping, electrical conductivities are several orders of magnitude smaller than those of doped $(CH)_x$. Apparently, substituents permit or induce twisting in the polymer backbone to relieve steric interactions and, as a result, the effective conjugation length is considerably reduced due to poor π-orbital overlap. Recently, significantly improved conductivity has been reported upon doping of poly(1,6-heptadiyne)[13]. In this case, the "substituent" is actually a ring which bridges the polymer chain and presumably helps to lock the chain in an approximately planar configuration.

An interesting post-polymerization reaction which affords a partially substituted polyacetylene has been reported by Kletter and coworkers[14]. They found that appropriate thermal treatment of bromine-doped $(CH)_x$ gave partial replacement of hydrogen by bromine on the polymer backbone. Subsequent doping of this material with I_2 or AsF_5 afforded conductivities two to three orders of magnitude lower than the corresponding doped $(CH)_x$.

An alternative approach to $(CH)_x$ derivatives is through copolymerization of acetylene with various monomers. Experimental control of physical properties through judicious choice of the comonomer composition is an attractive aspect of this method. It might be anticipated, for example, that the high conductivity characteristic of doped polyacetylene and the solubility of an appropriate polymer can be combined into one material through copolymerization. In addition, this approach may provide answers to fundamental questions concerning the dependence of conductivity on

the nature and the amount of defects introduced onto the
$(CH)_x$ backbone.

With the above considerations in mind, we initiated a study
of the feasibility of the copolymerization method in the production of novel conducting materials. The study focused on acetylene-methylacetylene copolymer films. Methylacetylene was selected as the comonomer for three reasons. First, the homopolymer is soluble in common organic solvents. Second, the small
methyl substituent was not expected to significantly disrupt interchain interactions, allowing electrical transport to be determined primarily by structural intrachain properties. Finally,
methylacetylene is a gas above ca. -25°C (1 atm.) and allows the
use of techniques developed for $(CH)_x$ film synthesis to be
applied to copolymer film preparations. The preparation of films
was considered to be important in order to more easily compare
electrical properties to those of the rather well-characterized
$(CH)_x$ films.

It is important to note that Wegner et al.[15] recently reported the synthesis of copolymers of acetylene and 1-hexyne.
These workers found that λ_{max} increased (i.e., the band gap decreased) with increasing acetylene content in the copolymers.
However, chemical doping of these materials has not yet been investigated.

Experimental

Equipment and Purification of Materials. Toluene was repeatedly washed with concentrated H_2SO_4, distilled H_2O, aqueous 10%
NaOH and additional distilled H_2O followed by drying over anhydrous
$MgSO_4$ and refluxing over CaH_2 under argon. The toluene was further dried with Et_3Al and was distilled into the polymerization
reactor on a vacuum line.

The polymerization reactor (essentially a Schlenk tube) was
constructed of glass; stopcocks employed Teflon plugs. The reactor was equipped with a sidearm for repeated washing of the copolymer films with freshly-distilled solvent.

The comonomer gas reservoir was a 2ℓ glass bulb containing
methylacetylene (Air Products, dried over P_2O_5 at -78°C) and
acetylene (Union Carbide, passed through two bubblers of concentrated H_2SO_4 and a U-tube containing P_2O_5). The total pressure of
the gas mixture was never greater than ca. 720 torr. The gases
were assumed to be ideal and the mole ratios were determined
simply by the partial pressures of the gases in the bulb. Traces
of air which may have entered the bulb during gas transfers were
removed by employing at least two freeze-pump-thaw cycles.

Copolymer Film Synthesis. Initial syntheses employed a Ziegler-Natta catalyst system initially used by Shirakawa for the production of $(CH)_x$ film, i.e. 1.7 ml of $Ti(OBu)_4$ and 2.7 ml of Et_3Al in 20 ml of toluene. Copolymerizations were carried out at -10°C (ice-salt bath) to preclude condensation of the methylacetylene fraction of the comonomer feed (b.p. of methylacetylene is ca. -23°C at 1 atm.). The viscosity of the catalyst solution at -10°C was rather low and, therefore, did not adhere to the reactor walls effectively. As a result, extremely thin (ca. 15-40 μm) copolymer films were obtained. In many cases, the film could not be removed from the reactor walls in workable pieces; rather, thin flakes were usually obtained.

Higher quality films were obtained using a more concentrated (and therefore more viscous) catalyst solution. This was prepared by adding 2.5 ml of $Ti(OBu)_4$ and 4.0 ml of Et_3Al to 10 ml of toluene at 0°C followed by aging at room temperature for ca. 30 minutes. Reaction times with the comonomer feed ranged from 20-40 minutes. The films were washed repeatedly with toluene (15-20 times) to remove catalyst and low molecular weight soluble copolymer fractions. The films were handled in a dry box or a glove bag. Copolymer compositions were determined from elemental analyses. A $(CH)_x$ film was also prepared using the catalyst system described above.

Preparation of Poly(methylacetylene), $(C_3H_4)_x$. A stirred solution of 1.7 ml of $Ti(OBu)_4$, 2.7 ml of Et_3Al and 20 ml of toluene was exposed to methylacetylene (2ℓ bulb, ca. 700 torr initial pressure) for 1-2 hours. Using Schlenk tube techniques, the reaction mixture was syringed into a solution of methanol/HCl to precipitate the polymer. The polymer was dissolved in toluene, filtered and then reprecipitated. After washing several times with anhydrous methanol, the polymer was dried by pumping overnight. The product was an orange, flaky powder.

Doping and Electrical Conductivity. Rectangular copolymer and $(CH)_x$ films were mounted on platinum wires in a glass apparatus with Electrodag. Doping with AsF_5 and iodine (Fisher) was carried out in vacuo to saturation conductivities. The AsF_5 was maintained at ca. -95°C during the doping process. Conductivities were measured using standard four-probe techniques. Dopant concentrations were determined by weight uptake.

In the case of pure $(C_3H_4)_x$, a small quantity of the polymer was dissolved in toluene and cast on a thin rectangular glass plate in a glove bag. Platinum wires were attached to the dried film with Electrodag.

Spectra. EPR spectra were obtained using a Varian E-9 X-
band spectrometer equipped with a dual microwave cavities and var-
iable-temperature accessory.

Results

Copolymer Compositions. Table 1 summarizes pertinent data
concerning feed compositions and chemical analyses of the copolymer films and relevant homopolymers. Copolymer compositions were
determined from C/H ratios. Three general statements can be made
concerning the data. First, the acetylene content of the copolymers increases with increasing acetylene content in the feed.
Second, the total C and H contents of all of the materials are
low. Third, the C/H ratios of $(CH)_x$ and $(C_3H_4)_x$ are considerably
greater than those anticipated for the "pure" homopolymers.
These last two results are disappointing since they cast serious
doubts about the validity of determining copolymer compositions
from C/H ratios. However, plotting the feed composition vs. copolymer composition (Figure 6) affords a reasonable relationship
with only a moderate amount of scatter. It is doubtful that this
relationship is merely fortuitous, and this suggests that the copolymer compositions are not grossly in error. It should be
noted, however, that since the C/H ratios of the homopolymers are
larger than the anticipated values, it is likely that the copolymers are somewhat more rich in acetylene content than the data
indicate.

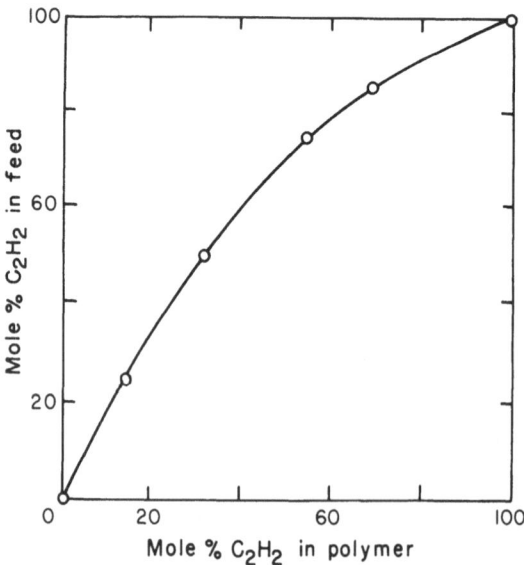

Figure 6: Composition of copolymer as function of mole %
           acetylene in feed.

Qualitative Comparisons. A $(CH)_x$ film was prepared using the catalyst system employed for copolymer film synthesis simply for comparison. The $(CH)_x$ film so prepared was quite different in appearance than the "typical" films. First, the film possessed poor integrity and broke into several small pieces during washing. Second, the film was very thin (ca. 20 μm) and extremely brittle, similar to that obtained using a neat (nonsolvent) catalyst solution. The film surfaces were very irregular and displayed a macroscopic "rippled" appearance. The same observations apply to copolymer AMA-61, (Table 1).

Sample AMA-31 more resembled typical $(CH)_x$ films, i.e., a shiny green side which grew against the glass reactor wall and a dull, grey-black backside. Electron micrographs showed no evidence of the fibrillar morphology characteristic of typical $(CH)_x$ films. Rather, irregular and ill-defined "clumps" were observed.

Copolymer AMA-11 displayed a dull greenish-gold luster on both sides of the film. The morphology was similar to that of AMA-31. The luster was lost and the film became black upon wetting with a solvent (e.g., pentane, toluene), suggesting that solvent swelling of the polymer occurred. The luster was recovered upon evaporation of the solvent. The AMA-11 film became elastic upon wetting with a solvent. Wetted films could be stretched to extension ratios as high as ~7 and, upon solvent evaporation, a fixed elongation could be achieved. Alternatively, wetting of a stretched, dried film resulted in considerable retraction.

The same general observations concerning sample AMA-11 apply to the characteristics of the AMA-13 film although the latter was less lustrous and even more elastic when wetted with solvent. An attempted preparation of an AMA-14 copolymer failed to produce a free-standing film. The material was extremely rubbery and tacky in the presence of solvent and, as the result of poor mechanical integrity, was washed away from the reactor walls as black, tacky masses. Thus, it appears that the methylacetylene content of AMA-13 was nearly the upper limit for the production of free-standing films.

Poly(methylacetylene), $(C_3H_4)_x$, was obtained as a brittle orange powder. This material is much more sensitive to oxidation than $(CH)_x$. The orange powder became yellow-white in color upon overnight exposure to oxygen suggesting that considerable disruption of the conjugated backbone occurred. In contrast, $(CH)_x$ lost its luster and became discolored only after several weeks of air exposure. Copolymer films rich in methylacetylene content (e.g., AMA-11, AMA-13) became yellow-white and extremely brittle merely upon overnight air exposure, resembling $(C_3H_4)_x$. It was also found that $(C_3H_4)_x$ is less thermally stable than $(CH)_x$; decomposition temperatures were ca. 200°C and ca. 320°C, respectively. The copolymers displayed intermediate behavior.

TABLE 1

Comonomer Feed and Copolymer Composition

| Sample[b] | Mole Ratio C_2H_2/C_3H_4 in Feed | Mole % C_2H_2 in Feed | Chemical Analysis of Polymers[a] | | | | Mole % C_2H_2 in Feed |
			% C	% H	Total C,H %	H/C	
$(CH)_x$	∞	100	84.16	7.61	91.77	1.08(1.00 theor.)	(100)
AMA-61	6	85	86.31	8.15	94.46	1.13	70
AMA-31	3	75	87.31	8.61	95.92	1.18	55
AMA-11	1	50	86.36	8.94	95.30	1.24	33
AMA-13	0.33	25	84.53	9.30	93.83	1.31	15
$(C_3H_4)_x$[c]	0	0	88.81	10.95	99.76	1.48(1.33 theor.)	(0)

a – Galbraith Laboratories, Inc., Knoxville, Tenn.

b – Code = Acetylene – MethylAcetylene – mole ratio C_2H_2/C_3H_4 in feed

c – Courtesy of Mr. J.M. Warakomski, this laboratory

All copolymer samples and (CH)$_x$ yielded Lorentzian EPR resonances with g values in the range 2.0023-2.0026. Line widths
ranged from ca. 10 G to ca. 7 G and in general increased with in-
increasing methylacetylene content in the copolymers, Figure 7.
Heating each sample to 150°C for ca. 30 minutes in the EPR probe,
followed by cooling to room temperature, resulted in the relationship shown. (C$_3$H$_4$)$_x$ did not display any observable resonance even
when heated to 150°C, although the material melted during heating.

Electrical Conductivity. Conductivities and dopant concentrations for (CH)$_x$, (C$_3$H$_4$)$_x$ and various copolymer samples are
given in Table 2. The conductivities decreased with decreasing
acetylene content, Figure 8. Iodine became a progressively better
dopant than AsF$_5$ as the methylacetylene content of the polymers
increased. The conductivity of the red, soluble residue obtained
from an AMA-11 preparation and doped with iodine was essentially
identical to a similarly doped sample of (C$_3$H$_4$)$_x$. Due to the
oily (low molecular weight) nature of this material and the low
conductivity, little effort was made to characterize the soluble
fractions.

The poor purity of the samples investigated (Table 1) is
probably the result of considerable catalyst entrapment in the

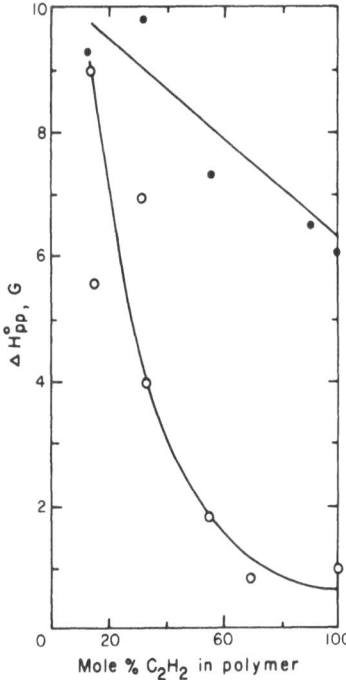

Figure 7: EPR linewidth as a function of copolymer composition:
○, ambient temperature; ○, after heating to 150°C, 30
mins., and return to ambient.

TABLE 2

Conductivities and Compositions of Doped Homo- and Copolymers

Sample[a]	Mole % C_2H_2 in Polymer	Composition of Doped Polymer	σ, Ω^{-1} cm^{-1}
$(CH)_x$	100	$[CH(AsF_5)_{0.12}]_x$	~400
AMA-31	~55	$[CH_{1.18}I_{0.24}]_x$	~36
		$[CH_{1.18}I_{0.16}]_x$ - after pumping above sample overnight	~18
		$[CH_{1.18}(AsF_5)_{0.08}]_x$	~45
AMA-11	~33	$[CH_{1.24}I_{0.16}]_x$	~1.5
		$[CH_{1.24}(AsF_5)_{0.1}]_x$	~1.0
AMA-13	~15	$[CH_{1.31}I_{0.11}]_x$	~2×10^{-2}
		$[CH_{1.31}(AsF_5)_{0.05}]_x$	~2×10^{-3}
$(C_3H_4)_x$[b]	0	$[CH_{1.33}I_{0.17}]_x$[b]	~10^{-3} [b]

a - Table 1

b - Reference 11

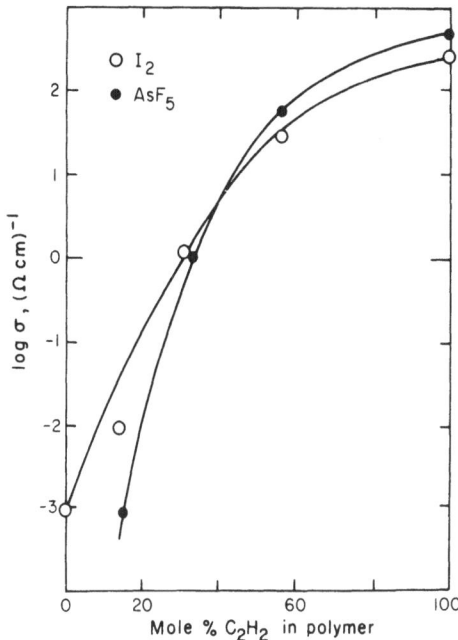

Figure 8:   Conductivity vs. copolymer composition for doped co-
            polymer films.

materials due to the high catalyst concentrations employed in
their preparation. This effect was most pronounced in the $(CH)_x$
sample since it is not swellable with solvents and thus an un-
usually large quantity of catalyst could remain in the material.
However, the conductivity of this AsF_5-doped $(CH)_x$ sample (Table
2) is comparable to that obtained with typical as-grown films.
Furthermore, this $(CH)_x$ sample displayed identical EPR behavior
as typical as-grown films (Figure 7); i.e., the ca. 6 G resonance
narrows to ca. 1 G upon isomerization. Thus, the purity of the
$(CH)_x$ appears to affect neither its "dopability" or its EPR behavior to any great degree and suggests that copolymer results
are not particularly dependent upon sample purity.

     The oxidative instability of $(C_3H_4)_x$ and the copolymers is a
direct consequence of the methyl substituent on the conjugated
backbone. These hydrogens are readily susceptible to radical
attack since this leads to resonance stabilized radicals.

     The qualitative results described previously suggest the
presence of true copolymers as opposed to "blends" of $(CH)_x$ and
$(C_3H_4)_x$. However, definitive proof of copolymer formation was
difficult to obtain because of the insolubility of the films.
The effective conjugation length of $(C_3H_4)_x$ is expected to be severely reduced as compared to $(CH)_x$ due to steric interactions between methyl and hydrogen substituents, causing a "twist" of the
backbone. In fact, a uv-visible spectrum of $(C_3H_4)_x$ in heptane

shows two λ_{max} absorptions at ca. 220 and 290 nm; these are of
higher energy compared to $(CH)_x$ (λ_{max} ca. 600-700 nm). In other
words, the band gap of $(C_3H_4)_x$ is considerably larger than that of
$(CH)_x$. As a result of these steric factors, the $(C_3H_4)_x$ polymer
backbone is less rigid and, taken in conjunction with the disruption of intermolecular interactions, the polymer is soluble.
Apparently, the methylacetylene content of copolymers AMA-11 and
AMA-13 is insufficient to induce a great deal of solubility
although interchain interactions are disrupted to the extent that
solvents can plasticize the materials.

The effective conjugation length of the copolymers will increase with increasing acetylene content as demonstrated by Wegner et al.[15] with acetylene/1-hexyne copolymers. UV-visible spectra for these copolymers are not yet available. However, the EPR
data of Figure 7 argue in favor of the presence of "true" copolymers. No EPR resonance for $(C_3H_4)_x$ was observed even when the
sample was heated to 150°C. EPR behavior from a polymer blend
would be characteristic of the highly conjugated $(CH)_x$ itself.
On the other hand, true copolymers are expected to possess reduced mobility of the unpaired spins in the polymers. The EPR
line-width in turn qualitatively reflects the mobility of the
spins. Figure 7 indicates that the line-width does increase (and
thus the mobility decreases) with increasing methylacetylene content in the polymers. This dependence is even more pronounced
with samples which have been isomerized by heating to 150°C.
These results suggest that samples AMA-31, AMA-11 and AMA-13 are
not simply mixtures of homopolymers. The absence of an observable
EPR resonance for $(C_3H_4)_x$ is difficult to rationalize since a resonance is observed in the related material, poly(phenylacetylene)[16]. It is interesting to note that the EPR line-width of
poly(phenylacetylene) is ca. 10-12 G; i.e., consistent with the
anticipated line-width of poly(methylacetylene) (Figure 7).

The electrical conductivity data of Table 2 and Figure 8 are
consistent with predictions concerning the dependence of conductivity on copolymer composition. An increase in the methylacetylene content of the polymer is expected to increase both its band
gap ($\pi \rightarrow \pi^*$ transition) and its ionization potential, making free
carrier generation more difficult. In addition, the reduction of
the effective conjugation length with increasing methylacetylene
content is expected to reduce carrier mobility. Thus, the methylacetylene units would act as barriers to conductivity. Carrier
migration through (or around) these barriers should become progressively more difficult as the number of these barriers increases.

As the acetylene content of the copolymers decreases, AsF_5
becomes a progressively poorer dopant compared to iodine. In
fact, AsF_5 produces no significant change in the conductivity of

$(C_3H_4)_x$. This phenomenon might be due to a complicated combination of steric and electronic factors with regard to polymer-dopant interactions. These interactions appear to be weak in the case of $(C_3H_4)_x$ doped with iodine. Most of the iodine can be removed from the sample by pumping; this produces a six orders of magnitude decrease in conductivity. Removal of iodine from doped copolymers by pumping resulted in a small decrease in conductivity (Table 2). This behavior is similar to that observed with iodine-doped $(CH)_x$.

Conclusions

The applicability of copolymerization in the synthesis of novel electrically conducting materials having interesting properties (i.e., solvent plasticization) has been demonstrated. The electrical conductivity of the doped copolymers increases with increasing acetylene content. It is suggested that the methylacetylene units in the polymers are the predominant barriers to electrical conductivity.

Detailed studies of the morphology, degree of crystallinity and thermopower of the copolymers are in progress. The latter is representative of intrinsic electrical properties and this parameter will be very useful in properly assessing the influence of copolymer composition on the electrical properties of these materials.

The acetylene-methylacetylene copolymers are not practical in view of their extreme sensitivity toward oxidation. However, the observation that partial methylation of the backbone can induce swellability is important and suggests the possibility of incorporating oxidative stabilizers into such materials. Upon consideration of the enormous number of potential copolymers which may be prepared, a large new class of organic materials is envisioned which possess desirable physical and mechanical properties and electrical conductivities which can be controlled over the full range from insulator to semiconductor to metal.

SECTION III: AROMATIC CONDUCTING POLYMERS

Introduction

The limitations of polyacetylene have been discussed in the preceding section. In contrast, the aromatic material poly(p-phenylene) exhibits excellent thermal and oxidative stability and gives a material with conductivity comparable to that of doped poly(acetylene) on doping with AsF_5. Doped poly(p-phenylene) is reasonably stable, but like the undoped polymer[17] is unprocessible. This has led us to investigate poly(p-phenylene vinylene), PPV, 1, which can be viewed as an alternating copolymer of "p-phenylene" and acetylene[18].

$$R' \left[-\left\langle \bigcirc \right\rangle -CH{=}CH- \right]_x -R^2$$

<u>a</u> R' = R² = CHO

<u>b</u> R' = CH₃ , R² = CH₂Cl

<u>I</u>

Synthesis and electrical properties of undoped PPV and numerous derivatives have been described in the literature[19],[20]. The wide variety of substituent groups which can be introduced on the aromatic ring offers the means for wide modification of processibility and electrical properties.

Synthesis

PPV was synthesized using the Wittig olefin synthesis as described by McDonald and Campbell[21]. A slight excess of terephthaldehyde ensured that the polymer possessed aldehyde end groups, confirmed by IR absorption at 1690 cm^{-1}, and allowed determination of the average number of repeat units, $\bar{x}$. The initial polymer, which contains both cis and trans vinylene units was isomerized to the all trans material by reflux in toluene with a catalytic amount of iodine. Elemental analyses for two samples, shown in Table 3, corresponded to values of $\bar{x} = 8 \pm 1$ and ~3. PPV was also prepared by dehydrochlorination of p-xylidene dichloride using the method of Horhold[20]. Elemental analysis for this material both before and after pyrolysis to eliminate HCl are given in Table 3. Substituted PPV analogs were prepared as described in the literature, Table 5.

Poly(p-phenylene vinylene)

Room temperature conductivity of a thin PPV wafer is ~10^{-9} (ohm cm)$^{-1}$, somewhat higher than the value reported by Manecke et al.[22]   Exposure of the pellet, attached to a 4-probe apparatus, to ~30 torr of the electron acceptor AsF$_5$ give a conductivity increase of ca. ten orders of magnitude in 2-3 hrs. The lemon yellow sample ultimately became dark brown with a brassy luster. The dependence of conductivity on uptake of dopant, determined by weight uptake of a reference PPV wafer, is shown in Table 4. A significant and surprising finding is that the limiting conductivity was essentially the same for the high molecular weight and low molecular weight samples.

PPV synthesized by dehydrochlorination of p-xylidene dichloride gave a similar increase in conductivity when exposed to arsenic pentafluoride. This demonstrates that the conductivity

TABLE 3

Elemental Analysis of Polymer Samples

Sample	Analysis, %[a]			Empirical Formula	$\bar{x}$
	C	H	O		
1a-1	90.6	5.74	3.40	$\sim C_{72}H_{54}O_2$	~8
1a-2	87.71	6.07	6.20	$\sim C_{32}H_{24}O_2$	~3
7	88.92	6.58	2.88	$\sim C_{108}H_{86}O_2$	~10
	C[a,b]	H[a,b]	Cl[a,b]		
1b - before pyrolysis	84.40	6.15	7.50	$C_{8.00}H_{6.94}Cl_{0.24}$	
1b - after pyrolysis	88.7	6.20	2.75	$C_{8.00}H_{6.70}Cl_{0.08}$	~11[c]

a. Galbraith Laboratories, Inc., Knoxville, TN

b. University of Massachusetts Microanalytical Laboratory

c. Assuming one - CH_2Cl group per chain

TABLE 4

Room Temperature Conductivity
of Arsenic Pentafluoride-Doped PPV[a]

$\sigma (ohm\ cm)^{-1}$	AsF_5 (wt. %)
4×10^{-5}	5.6
1×10^{-3}	8.9
6.2×10^{-2}	16.4
~3	57.0

[a] Wafer (~10 mil thickness) pressed
between Teflon sheets at ca. 300 psi

increase on AsF$_5$ doping is a property of PPV and is not caused by
trace impurities since these should be different for the two synthetic routes. It is significant that again limiting conductivity
does not appear to depend on the length of the conjugated system
present in PPV prior to doping. Vacuum pyrolysis at 300°C for 3
hr. of PPV synthesized by dehydrochlorination results in a ca. 8%
weight loss and a color change from yellow to deep orange. Elemental analyses (Table 3) are consistent with loss of HCl being
the major process. Assuming one chloromethyl terminal per chain,
analysis after pyrolysis is consistent with x = 11 ± 1. Analytical
data obtained before pyrolysis are consistent with a chain of 11
units containing an average of four chlorine atoms which would
break the conjugated system at three sites, 2.

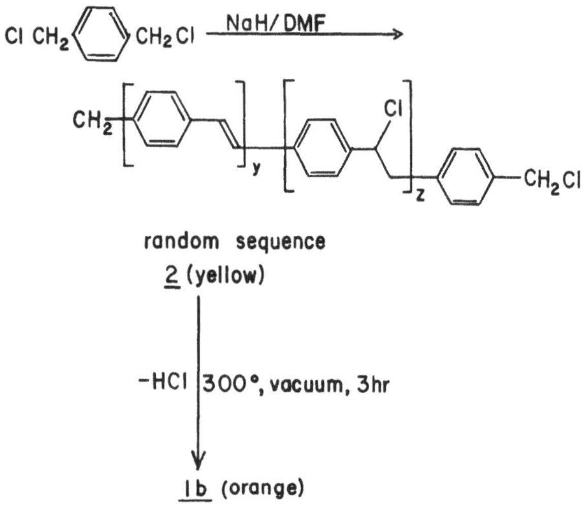

random sequence
2 (yellow)

−HCl 300°, vacuum, 3hr

1b (orange)

Nevertheless, 1b and 2 exhibit similar limiting conductivities on
AsF$_5$ doping with that of the less conjugated 2 higher by a factor
of 7.

     Arsenic pentafluoride-doped PPV exhibits many similarities
to doped poly(acetylene) and poly(p-phenylene). As in the latter,
conductivity of PPV is not increased by exposure to iodine or
ammonia vapors. Conductivity of arsenic pentafluoride-doped PPV
is electronic; passage of an 0.5 mA dc-current through a sample
for 14h caused almost no change in resistance. The chemical compensation of donor and acceptor dopants observed for poly(acetylene) and poly(p-phenylene) occurs in doped PPV as well. Exposure
of arsenic pentafluoride-doped PPV [σ = 3 (ohm cm)$^{-1}$] to ca. 100
torr of ammonia resulted in a rapid decrease in conductivity and
gave a light brown insulator after several minutes.

     Metal-like properties were apparent in doped PPV. The IR
spectrum of pure PPV pressed on Teflon film (with Teflon film in

the reference beam of the spectrometer) is shown in Figure 9 (full curve). Exposure of this sample to AsF_5 at 30 torr for 3 hrs. rendered the spectrum featureless (broken curve), the high transmission being caused by passage of radiation through the Teflon film owing to incomplete coverage by the PPV sample. The featureless spectrum is consistent with metallic behavior. Electron paramagnetic resonance has been used to study free electron spins and the effect of doping. Undoped PPV exhibits no EPR signal, but exposure to ca. 1 torr arsenic pentafluoride, for 1 min. resulted in a narrow (width 0.5 G) signal (Figure 10A), presumably the result of oxidation by arsenic pentafluoride. Heavily doped PPV exhibited a broadened signal (width 1.8 G) which is asymmetric (Figure 10B). The Dysonian lineshape, which has been observed for AsF_5-doped poly(acetylene), is consistent with metallic behavior in AsF_5-doped PPV. Finally, in a preliminary study of temperature dependence of conductivity in AsF_5-doped PPV we have found that the dependence decreases markedly as the dopant level increases. Heavily-doped samples of limiting conductivity have not yet been studied, but a sample with ca. 15 weight percent arsenic pentafluoride suffered less than a ten-fold conductivity decrease on going from room temperature to 77°K.

Table 5 presents conductivity data for PPV samples as well as several low molecular weight PPV models and some PPV analogs. The low molecular weight models, 3, 4 and 5, exhibit a trend of increasing conductivity with length of the conjugated system that is not manifested by the more highly conjugated PPV's. Planarity and conjugation in doped PPV may be important however since the mixed cis and trans isomer which is likely to be non-planar, and poly(m-phenylene vinylene), which is a white material with little conjugation, both exhibit reduced conductivity. The methoxylated PPV, 8, was designed to have a lower ionization potential than

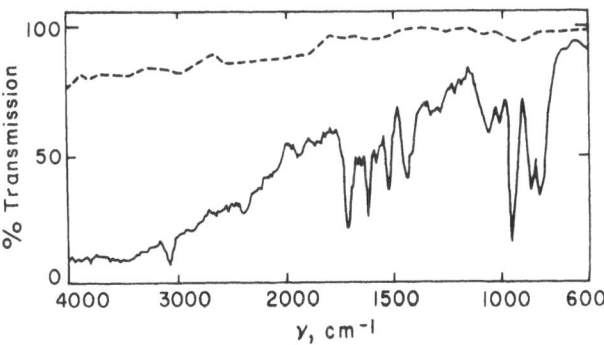

Figure 9:   IR spectra of undoped PPV (full line) and
                AsF_5-doped (broken line).

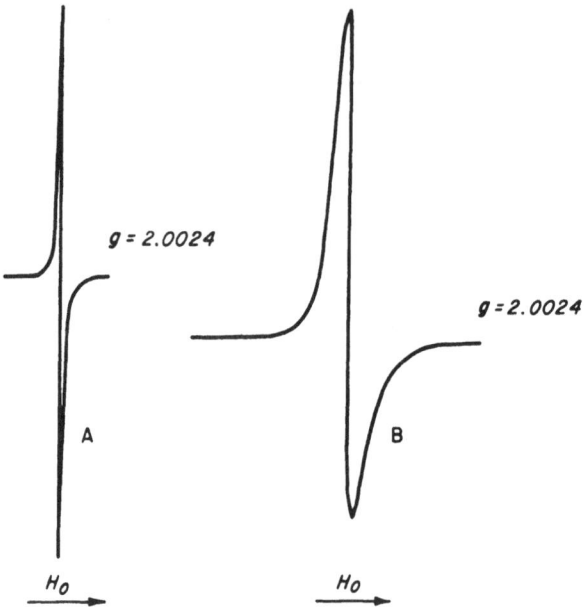

Figure 10:   EPR spectra of AsF$_5$-doped PPV.   A: lightly doped;
             B: heavily doped.[5]

PPV in the hope that it would be doped by milder oxidants than
AsF$_5$.   It was not doped by I$_2$,[2] and its minimal increase on conductivity when treated with AsF$_5$ may be caused by coordination
of the strong Lewis acid AsF$_5$ at[5] the basic methoxy oxygens which
would elevate the ionization[5] potential of the polymer making creation of charge carriers by oxidation more difficult.

   In all of the above cases in which pressed pellets were
doped with AsF$_5$ vapor,[2] only a thin layer at the surface of the
sample was doped.   Fracture of the doped pellets revealed light-
colored undoped interiors.   Two implications are obvious.   First,
since conductivities were computed on the assumption that the entire pellet conducts, the actual conductivities of the fully-doped
materials are higher than those reported.   Second, some change
must occur on doping which forms a "skin" that prevents diffusion
of AsF$_5$[2] into the interior of the pellet.   A logical possibility
is further polymerization of the starting polymer via Friedel-
Crafts type chemistry catalyzed by AsF$_5$.   Our observation that
the limiting conductivity of AsF$_5$-doped PPV is independent of the
average degree of polymerization[5] ($\bar{x}$) of the starting polymer may
be due to this phenomenon.

TABLE 5

Conductivity of Arsenic Pentafluoride-Doped PPV and PPV Analogs

Compound	Conductivity (ohm cm^{-1})^{a,b} Undoped	Doped	Wt.% AsF$_5$
Ph—CH=CH—Ph **3**	$< 10^{-9}$	ca. 10^{-6}	18
structure **4**^{c}	$< 10^{-9}$	ca. 10^{-5}	11
structure **5**^{d}	$< 10^{-8}$	ca. 10^{-4}	16
PPV **1a** cis and trans	$< 10^{-9}$	ca. 10^{-2}	--
PPV **1a** all trans	$< 10^{-9}$	3	57
PPV **1b** all trans	$< 10^{-9}$	0.7	18
PPV **2**	$< 10^{-9}$	ca. 10^{-1}	30
structure **6**^{e}	$< 10^{-9}$	$< 10^{-7}$	--

TABLE 5 continued

Compound		Conductivity (ohm cm^{-1})[a,b]		Wt.% AsF$_5$
		Undoped	Doped	
	7[f]	< 10^{-8}	ca. 10^{-3}	50
	8[g]	< 10^{-9}	ca. 10^{-7}	80

a. Conductivities below 10^{-4} (ohm cm)$^{-1}$ measured by the 2-probe technique. Conductivities above 10^{-4} measured by 4-probe technique.

b. With the exception of PPV 1a all trans, pellets were pressed in evacuated KBr dies at ca. 10,000 psi. Conductivities were calculated assuming homogeneous doping and conduction throughout the pellet (see text).

c. G. Kossmehl, Ber. Bunsenges. Phys. Chem., 83 417 (1978) and S. Misumi et al., Bull. Chem. Soc. Japan., 34, 1833 (1961).

d. S. Misumi, et al., Bull. Chem. Soc. Japan, 36, 399 (1963).

e. G. Kossmehl, M. Hartel and G. Manecke, Die Makromol. Chem., 131, 37 (1970).

f. G. Kossmehl, Ber. Bunsenges. Phys. Chem., 83, 417 (1978).

g. G. Manecke, D. Zerpner and G. Kossmehl, Die Makromol. Chem., 147, 35 (1971).

Acknowledgements

The authors wish to acknowledge support from UMass. Materials Research Laboratory, ONR, and DARPA.

References

1.   K. Soga, S. Kawakami, H. Shirakawa and S. Ikeda, Makro.
      Chem., Rapid Comm., 1, 523 (1980.
2.   V. Chacko, J.C.W. Chien, F.E. Karasz, A. Heeger and A. Mac-
      Diarmid, Bull. Am. Phys. Soc. 24, 480 (1979).
3.   C.F. Feldman and E. Perry, J. Polym. Sci. 46, 217 (1960).
4.   D.R. Burfield and P.J.T. Tait, Polymer 13, 315 (1972).
5.   G.D. Bukatov, N.B. Chumaevskii, V.A. Zakharov, G.I. Kuznet-
      sova and Y.I. Yermakow, Makromol. Chem. 178, 953 (1977).
6.   C.K. Chiang, M.A. Druy, S.C. Gau, A.J. Heeger, E.J. Louis,
      A.G. MacDiarmid, Y.W. Park and H. Shirakawa, J. Amer. Chem.
      Soc. 100, 1013 (1978).
7.   C.K. Chiang, S.C. Gau, C.R. Fincher, Jr., Y.W. Park, A.G.
      MacDiarmid and A.J. Heeger, Appl. Phys. Lett. 33, 18 (1978).
8.   D.M. Ivory, G.G. Miller, J.M. Sowa, L.W. Shacklette, R.R.
      Chance and R.H. Baughman, J. Chem. Phys. 71, 1506 (1979).
9.   K.K. Kanazawa, A.F. Diaz, R.H. Geiss, W.D. Gill, J.F. Kwak,
      J.A. Logan, J.F. Rabolt and G.B. Street, J. Chem. Soc. Chem.
      Comm. 854 (1979).
10.   T. Yamamoto, K. Sanechika and A. Yamamoto, J. Polym. Sci.,
      Polym. Lett. Ed. 18, 9 (1980).
11.   Y. Matsumura, A.G. MacDiarmid and A.J. Heeger, private
      communication.
12.   P. Cukor, J.I. Krugler and M.F. Rubner, ACS Polymer Pre-
      prints, 21, 161 (1980).
13.   H.W. Gibson, F.C. Bailey and J.M. Pochan, Org. Coat. Plast.
      Chem. 42, 603 (1980).
14.   M.J. Kletter, T. Woerner, A. Pron, A.G. MacDiarmid, A.J.
      Heeger and Y.W. Park, J. Chem. Soc. Chem. Comm. 426 (1980).
15.   V. Enkelmann, H. Muller and G. Wegner, Synthetic Metals 1,
      185 (1979/80).
16.   G.M. Holob, P. Ehrlich and R.D. Allendoerfer, Macromolecules
      5, 569 (1972).
17.   R.H. Baughman, D.M. Ivory, G.G. Miller, L.W. Shacklette and
      R.R. Chance, A.C.S. Preprint, Div. of Org. Coat. & Plastics,
      Washington, D.C., Sept. 1979, 41, 139.
18.   G. Wnek, J.C.W. Chien, F.E. Karasz and C.P. Lillya, Polymer
      20, 1441 (1979).
19.   G. Manecke, 24th. Int. Cong. of Pure & Appl. Chem., Hamburg,
      1973, 1, 155. M. Hartel, G. Kossmehl, G. Manecke, W. Wille,
      D. Wohrle and D. Zerpner, Die Angew. Makromol. Chem. 29/30,
      307 (1973); G. Kossmehl, Ber. Bunsenges. Phys. Chem. 83, 417
      (1979).

20. G. Drefahl, R. Kuhnstedt, H. Oswald and H.-H. Horhold, Die
     Makromol. Chem. 131, 89 (1970); H.-H. Horhold, J. Gottschaldt
     and J. Opfermann, J. Prakt. Chem. 319, 611 (1977).
21. R.N. McDonald and T.W. Campbell, J. Amer. Chem. Soc. 1960,
     82, 4669.
22. G. Kossmehl, M. Hartel and G. Manecke, Makromol. Chem. 1970,
     131, 37.

COFACIAL ASSEMBLY OF METALLOMACROCYCLES: A MOLECULAR

ENGINEERING APPROACH TO ELECTRICALLY CONDUCTIVE POLYMERS

Carl W. Dirk, Karl F. Schoch, Jr., and Tobin J. Marks

Department of Chemistry and the Materials Research
Center, Northwestern University, Evanston, Illinois
60201

INTRODUCTION

In the past few years, the new condensed matter chemistry/-
physics interface field of electronic materials having low-dimensional properties has undergone dramatic development.[1-6] The
preparation of advanced new classes of organic, metal-organic, and
inorganic substances with the properties of restricted dimension-
ality metals has excited theorists and experimentalists alike.
This activity is stimulating important breakthroughs in chemical
synthetic strategy and methodology, in spectroscopic, structural,
and transport analysis, and in the fundamental theoretical descrip-
tions of cooperative phenomena in the solid state. In terms of
technology, applications as varied as rectifiers,[7] sensors,[8,9]
solar energy conversion elements,[10,11] fuel cell components,[12]
switching devices,[13] photoresist elements,[14] chemoselective electrodes,[15,16] electrophotographic devices,[17] and durable synthetic
materials to replace metals[17] are being discussed. Speculation on
the possibility of high temperature superconductivity remains, as
always, a pervasive yet highly speculative motivation.[18,19]

Despite the intense activity in this field, our understanding
of the fundamental structural and electronic characteristics which
govern charge transport is at an embarrassingly primitive level.
This is equally true of the synthetic chemistry presently needed
to tailor molecular assemblies for evaluating contemporary theories
as well as for optimizing materials performance and processing
characteristics. The central focus of our research program has
been the development of rational syntheses of new low-dimensional
electronic materials and physical studies of the products arising
from this effort.[20,21] The goal has been to devise controllable

and systematic chemical approaches to such substances, so that
critical characteristics can be "tuned" and collective properties
correlated with changes in stoichiometry, crystal and electronic
structure, etc. Our initial efforts in this direction involved
materials composed of simple molecular stacks. However, as we
sought greater control over lattice architecture and asked more
fundamental questions about the molecular metallic state, the
structural inadequacies of such materials became readily apparent.

The purpose of this article is to review a new, successful
approach to the control of lattice architecture in low-dimensional
materials. It involves covalently linking and bringing into the
appropriate oxidation state, well-characterized and chemically
flexible molecular subunits. This approach capitalizes upon a
great deal of accumulated knowledge, and offers the possibility of
assembling a wide variety of robust, electrically conductive polymers with well-defined and easily modified primary and secondary
structures.[22-25] We begin with a general discussion of those factors promoting the molecular metallic state. Next we show how
this information and a cofacial assembly strategy lead to new
types of conductive polymers. Illustrating with the phthalocyanine subunit, the chemical and physical properties of the "face-
to-face" polymers are examined as a function of structure and
added dopants.

REQUISITES FOR CONDUCTIVE MOLECULAR MATERIALS

Two general features now appear to be necessary for converting
an unorganized collection of molecules into a conductive molecular
array.[1-6,20,21] First, the component molecules must be positioned
in close spatial proximity, with sufficient intermolecular orbital
overlap to provide a continuous electronic pathway for carrier
delocalization, and in crystallographically similar environments.
With the molecules positioned in this manner, the conduction pathway has a minimum of energetic "hills" and "valleys." Second, the
arrayed molecules must exist in formal fractional oxidation states
("mixed valence," "incomplete charge transfer," "partial oxidation").
In other words, the molecular entities to be connected in series
must have fractionally occupied electronic valence shells. Within
the framework of a simple one-dimensional Hubbard model, this prerequisite reflects the relatively narrow bandwidths (4t) and large
on-site coulomb repulsions (U) in such systems.[1-6] A simplified
valence bond picture of this situation is illustrated in Figure 1;
partial oxidation facilitates charge mobility by creating numerous
electronic vacancies. An analogous decription can likewise be
generated for partial reduction.

Our initial, first-generation strategy for the synthesis of
mixed valent low-dimensional materials involved the cocrystallization of planar, conjugated metallomacrocyclic donor molecules (D)

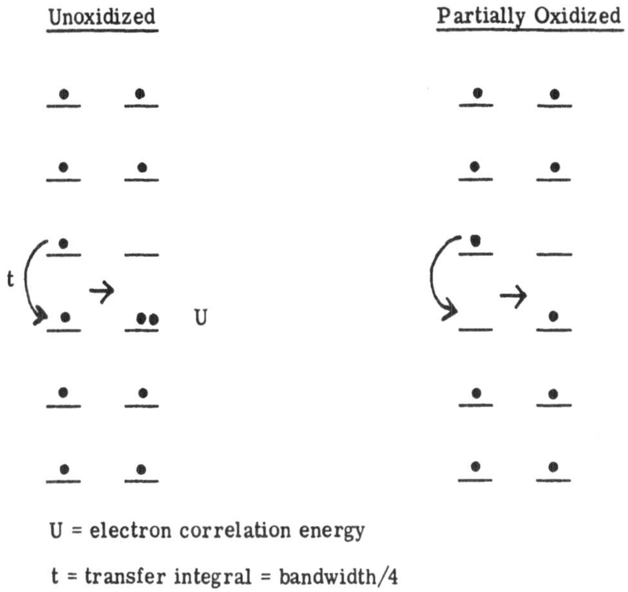
Unoxidized
Partially Oxidized
t
U
U = electron correlation energy
t = transfer integral = bandwidth/4

Fig. 1. Schematic illustration of the effect of partial oxidation on charge mobility in a low-dimensional system composed of molecular stacks.

having an MN_4 ligation sphere, with halogen acceptors (A) as shown below.[20,21] In optimum cases, the result has been lattices composed of segregated, partially oxidized metallomacrocyclic stacks and parallel arrays of halide or polyhalide counterions. An

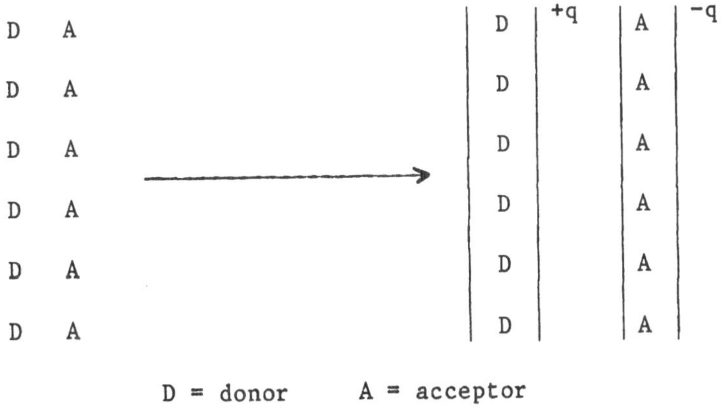
D A
D A
D A
D A
D A
D A
D
D
D
D
D
D
+q
A
A
A
A
A
A
-q
D = donor
A = acceptor

important additional feature of this approach is that the form of
the halogen (even if disordered) can be determined in a straight-
forward manner by resonance Raman/iodine Mössbauer spectroscopic
techniques.[21,26,27] The degree of partial oxidation follows from
this information and knowledge of the stoichiometry. This synthetic approach has enjoyed success for metal glyoximates,[28-32]
dibenzotetraazaannulenes,[33,34] phthalocyanines,[35-37] and por-
phyrins.[38] As an illustration, nickel phthalocyanine iodide,
$[Ni(Pc)]I_{1.0}$, consists of stacks of staggered $Ni(Pc)^{+0.33}$ units
arrayed at 3.244(2) Å intervals and surrounded by parallel chains
of I_3^- counterions. The 300°K conductivity of this material in
the molecular stacking direction is 300-700 Ω^{-1} cm^{-1} and the temperature dependence is "metal-like" ($\rho \sim T^{1.9}$) down to 60°K.[37]
The conductivity is predominantly a ligand-centered phenomenon,
and carrier mean free paths are comparable to some of the most
conductive "molecular metals."

THE COFACIAL ASSEMBLY STRATEGY

    Although the above molecular macrocycle, halogen cocrystalli-
zation approach to the synthesis of mixed valent low-dimensional,

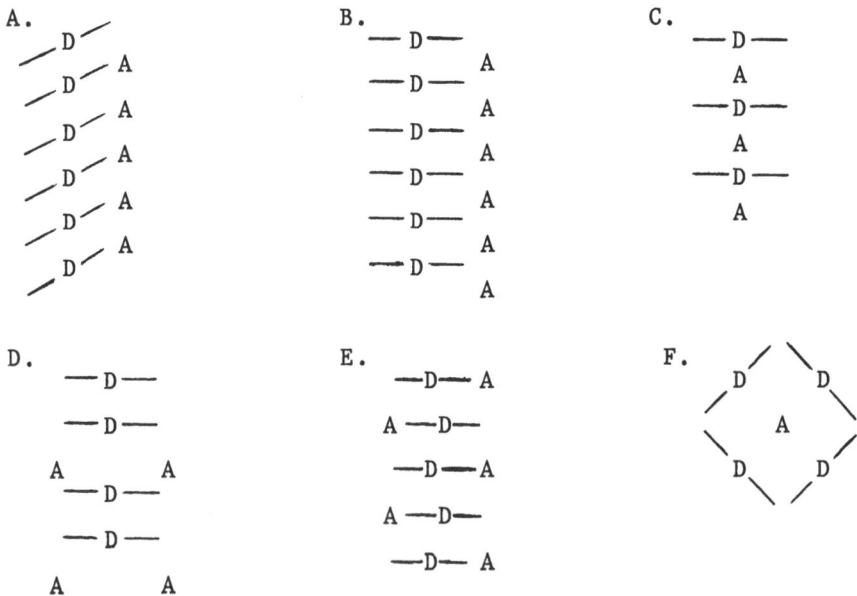

Fig. 2. Schematic illustration of some common structures for
         donor-acceptor complexes. A. Segregated stacking, canted
         donors. B. Segregated stacking, D_{nh} donor stacking. C.
         Integrated stacking. D. Integrated stacking, donor dimers.
         E. Segregated stacking, zig-zag donor stacking. F. Ion
         clusters without stacking.

metal-like materials is effective in many cases, it suffers, as do all strategies that rely upon molecular stacking, from a number of weaknesses. For example, in such cases the lattice architecture is dependent upon the unpredictable and largely uncontrollable forces that dictate the stacking pattern, the donor-acceptor orientations, and the stacking repeat distances. Figure 2 illustrates the complexity of the structural problem by depicting some of the types of common donor-acceptor crystallization patterns.[39-44] A frequent pitfall in the design of new materials is that segregated stacks (Figure 2A,B) do not form and that the effort expended in donor or acceptor design is for naught. For example, our efforts to substitute various oxidizing quinones for halogens in the cocrystallization synthesis have failed because integrated stack (Figure 2C,D) insulators are apparently formed.[45,46]

To achieve far greater control over molecular stacking we have employed the assembly process depicted in eq.(1).[22-25] Cofacial

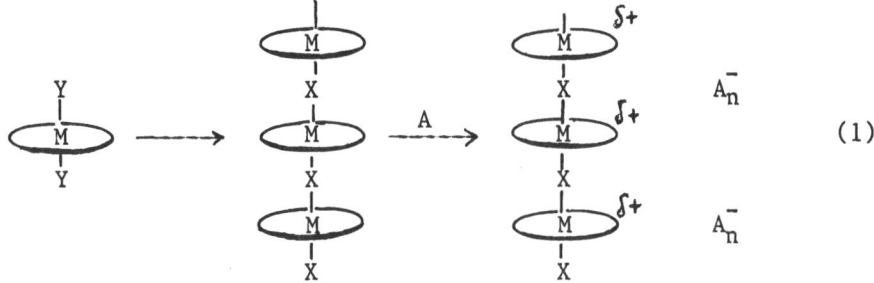

$$(1)$$

joining is carried out through the formation of strong covalent bonds involving the metallomacrocycle axial substituents (Y). The inherent variability of the metal, macrocycle, and bridging group offers great flexibility in terms of the structures which can be assembled, while the center-to-center mode of joining offers, in contrast to other conceivable modes of linking, superior structural control. Doping can then be carried out using either established halogenation methodology[20,21] or, as a consequence of the enforced stacking, entirely new types of dopants (vide infra).

Our first studies have involved phthalocyanine systems where M = Si, Ge, Sn, and X = O. Thus, $Si(Pc)(OH)_2$, $Ge(Pc)(OH)_2$, and $Sn(Pc)(OH)_2$[47,48] (we have developed an improved synthesis for the Si derivative) can be condensed at $300-400°C/10^{-3}$ torr to form "face-to-face" polymers (Figure 3). Concurrent to doping and transport studies, we have investigated the nature of these materials as polymers. The $[M(Pc)O]_n$ materials have high chemical and thermal stability; moreover, they are not significantly degraded by oxyen or moisture. We find that the polysiloxane polymer can be dissolved in concentrated sulfuric acid and recovered unchanged[49] (typical of phthalocyanines containing non-electropositive metals[50,51]). A rough estimate of the minimum average chain length

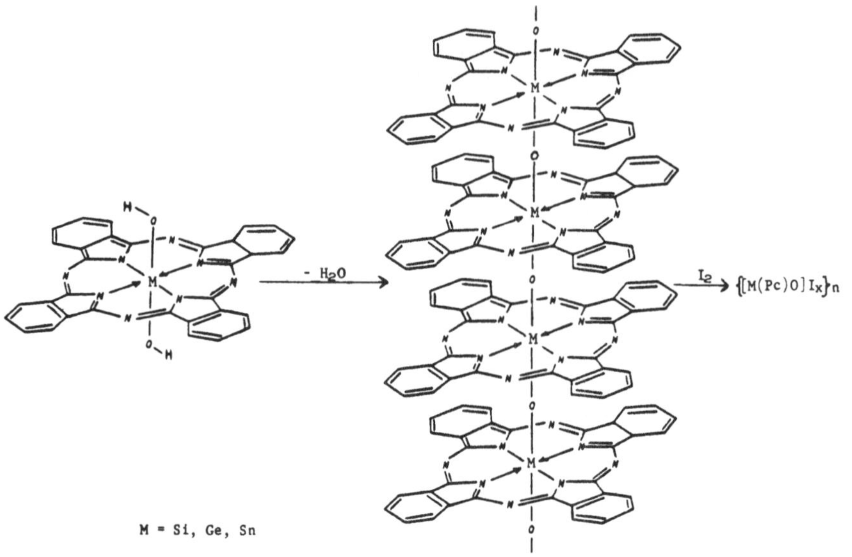

Fig. 3   Condensation reaction to produce cofacial arrays of Group
          IV metallophthalocyanines.

of $[Si(Pc)O]_n$ produced in the condensation polymerization can be
obtained by Fourier transform infrared spectrophotometric analysis
of the Si-O stretching region. For a typical sample, the degree
of polymerization is estimated to be on the order of ca. 100 subunits or more.[52] Prelimiliminary light scattering molecular weight
determinations in sulfuric acid solution are in agreement with
this result.[53]

Structural information on the face-to-face polymers has been
derived from several lines of evidence. We have indexed the X-ray
powder diffraction data in the tetragonal crystal system using
iterative computer techniques.[22-25] Data are very similar to
those from the columnar crystal structures of $[Ni(Pc)]I_{1.0}$[37] and
$Ni(dpg)_2I_{1.0}$.[29] The interplanar spacings in these latter tetragonal structures, determined in single crystal studies, are 3.244(2)
Å and 3.27(1) Å, respectively. The corresponding spacings derived
for the $[M(Pc)O]_n$ materials from the powder diffraction data
depend upon the ionic radius of the Group IV ion and vary from
3.33(2) Å (Si-O-Si), to 3.51(2) Å (Ge-O-Ge), to 3.95(2) Å
(Sn-O-Sn). The reliability of these metrical parameters is
further supported by single crystal diffraction results on the
model trimer $[(CH_3)_3SiO]_2(CH_3)SiO[Si(Pc)O]_3OSi(CH_3)[OSi(CH_3)_3]_2$
which contains three cofacial Si(Pc) units linked by linear
Si-O-Si connections at a distance of 3.324(2) Å.[54] In addition,
the $[Ge(Pc)O]_n$ and $[Sn(Pc)O]_n$ interplanar spacings obtained from
diffraction data are in good agreement with values we estimate
from ionic radii assuming linear Ge-O-Ge and Sn-O-Sn vectors,
i.e., 3.58 Å for $[Ge(Pc)O]_n$ and 3.90 Å for $[Sn(Pc)O]_n$.[22-25]

There is ample precedent for molecules with linear Si-O-Si, Ge-O-Ge, and Sn-O-Sn linkages.[55-58]

HALOGEN DOPING OF COFACIALLY LINKED METALLOPHTHALOCYANINES

Partial oxidation of the $[M(Pc)O]_n$ materials was first carried out using iodination methodology developed in this Laboratory for simple stacked systems. Stirring the powdered polymers with solutions of iodine in aromatic solvents or exposing the powders to iodine vapor results in a substantial iodine uptake. Alternatively, the $[Si(Pc)O]_n$ polymer can be doped by dissolving in sulfuric acid and precipitating with an aqueous I_3^- solution. The stoichiometries which can be obtained depend upon the reaction conditions; representative materials characterized by elemental analysis are compiled in the lefthand column of Table 1. A survey experiment also indicated that a bromine-doped material could be prepared. X-ray powder diffraction studies demonstrate that no major structural change in the face-to-face stack occurs upon halogenation.

Table 1. Physical Data for Polycrystalline Samples of Halogen-Doped $[M(Pc)O]_n$ Polymers

Compound	$\sigma(\Omega^{-1}cm^{-1})300°K$	Activation Energy (eV)	Interplanar Spacing (Å)
$[Si(Pc)O]_n$	3×10^{-8}		3.33(2)
$\{[Si(Pc)O]I_{0.50}\}_n$	2×10^{-2}		
$\{[Si(Pc)O]I_{1.55}\}_n$	1.4	0.04±0.001	3.33(2)
$\{[Si(Pc)O]I_{4.60}\}_n$	1×10^{-2}		
$\{[Si(Pc)O]Br_{1.00}\}_n$	6×10^{-2}		
$[Ge(Pc)O]_n$	$<10^{-8}$		3.51(2)
$\{[Ge(Pc)O]I_{0.31}\}_n$	7×10^{-4}	0.08±0.001	
$\{[Ge(Pc)O]I_{0.62}\}_n$	1×10^{-2}	0.05±0.001	
$\{[Ge(Pc)O]I_{1.94}\}_n$	6×10^{-2}	0.05±0.007	
$\{[Ge(Pc)O]I_{2.0}\}_n$	1×10^{-1}		
$[Sn(Pc)O]_n$	$<10^{-8}$		3.95(2)
$\{[Sn(Pc)O]I_{1.2}\}_n$	1×10^{-6}		3.95(2)
$\{[Sn(Pc)O]I_{5.5}\}_n$	2×10^{-4}	0.68±0.01	
$[Ni(Pc)]I_{1.0}$[a]	7×10^{-1}	0.036±0.001	3.244(2)

[a] Reference 35.

That oxidation of the cofacial array has indeed occurred is confirmed by resonance Raman scattering spectroscopy in the polyiodide region (Figure 4) which reveals the characteristic totally
symmetric stretching transition of I_3^- (ν = 108 cm^{-1}) and an
accompanying overtone progression. For stoichiometries with I/M
< 3, there are no more than traces of I_5^- ($\nu \approx$ 160 cm^{-1}) and no
evidence of free I_2 ($\nu \approx$ 200 cm^{-1}).[21] The nature of the doped
$[M(Pc)O^{\delta+}]_n$ electronic structure was also probed by electron spin
resonance (ESR). The symmetry of the lineshapes and the measured
g-values are consistent with π-radical cations, i.e., the unpaired
spin density is in molecular orbitals which are predominantly
ligand in character.[25] A similar conclusion was reached for
$[Ni(Pc)]I_{1.0}$.[37] ESR data for the $\{[M(Pc)O]I_x\}_n$ materials are set
out in Table 2.

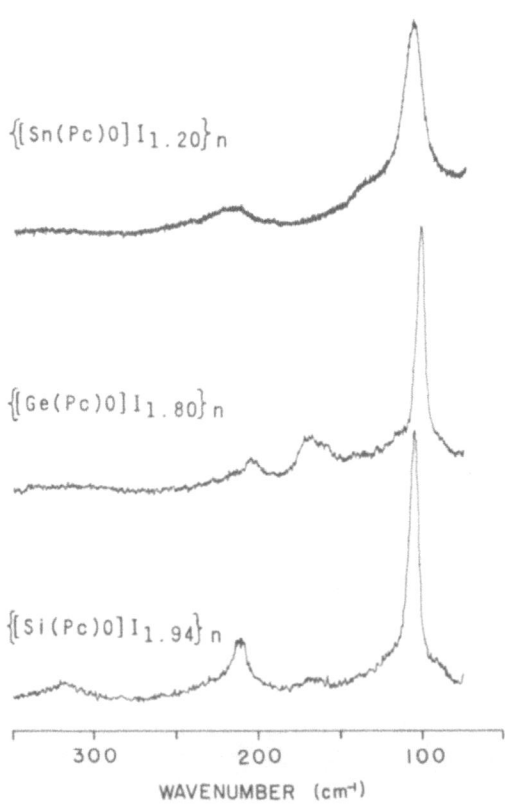

$\{[Sn(Pc)O]I_{1.20}\}_n$

$\{[Ge(Pc)O]I_{1.80}\}_n$

$\{[Si(Pc)O]I_{1.94}\}_n$

300   200   100

WAVENUMBER (cm^{-1})

Fig. 4.   Resonance Raman spectra (ν_0 = 5145 Å) of iodine-doped
           phthalocyanine face-to-face polymers. From ref. 22.

Table 2.   Powder ESR Data for Iodinated Phthalocyanine Face-to-
           Face Polymers

Compound	$g(300°K)^a$	$\Gamma(300°K)9G)^b$
$\{[Si(Pc)0]I_{1.40}\}_n$	2.003	5.1
$\{[Ge(Pc)0]I_{0.62}\}_n$	2.002	3.2
$\{[Sn(Pc)0]I_{1.20}\}_n$	2.002	6.0

a
 Average g-value; $g_{\parallel}$ and $g_{\perp}$ are not resolved.
b
 Observed linewidth.

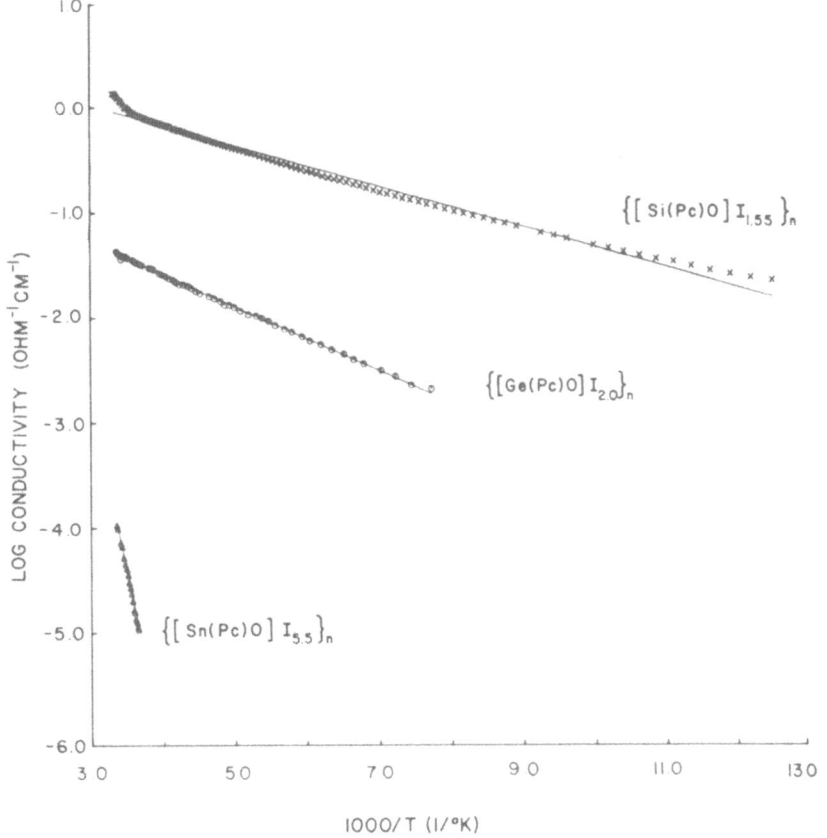

Fig. 5.   Variable temperature four-probe electrical conductivity
          data for powders of the iodine-doped, face-to-face phtha-
          locyanine polymers. The straight lines give least-
          squares fits to eq. (2). From ref. 24.

Four-probe van der Pauw[59,60] electrical conductivity measurements on the $[M(Pc)O]_n$ powders show them to be insulators. However,
iodine or bromine doping results in substantial increases in electrical conductivity (Table 1). The general trend in conductivity
as a function of metal is $\sigma_{Si} \gtrsim \sigma_{Ge} > \sigma_{Sn}$. Since it is known that
the transport characteristics of iodine-oxidized metallophthalo-
cyanines are largely ligand-dominated and relatively insensitive
to the identity of the metal,[35-37] the metal dependence of the
conductivity observed in the face-to-face polymers is logically
ascribed to structural differences such as how the interplanar
separation is influenced by metal ionic radius. Indeed, the
$\{[Si(Pc)O]I_x\}_n$ interplanar separation is within 0.1 Å of that in
the aforementioned "molecular metal" $[Ni(Pc)]_{1.0}$ and the room temperature powder conductivities of the two materials are quite comparable (Table 1). Variable temperature studies indicate that the
$\{[M(Pc)O]I_x\}_n$ powder conductivities are thermally activated
(Figure 5) and least-squares fits to eq. (2) yield the activation

$$\sigma = \sigma_0 e^{-\Delta/kT} \tag{2}$$

energies shown in Table 1. Powder conductivity measurements are,
of course, influenced by interparticle contact resistance and
averaging over all crystallographic orientations. Thus, for low-
dimensional compounds such as $[Ni(Pc)]I_{1.0}$, powder conductivities
are typically 10^2-10^3 less than single crystal conductivities in
the stacking direction and exhibit the thermally activated temperature dependencies. Thus, "metal-like" temperature dependence
($d\sigma/dT < 0$) is usually masked. From the powder conductivity data
on the $\{[M(Pc)O]I_x\}_n$ materials it is thus reasonable to anticipate that "metal-like" charge transport will be observed in the
chain direction for the M = Si and possibly M = Ge materials.
Further information on this question is provided by voltage
shorted compaction (VSC) techniques[61] which offer a qualitative
means to sample stacking axis transport properties in pressed
powder samples by deliberately shorting out sources of interparticle resistance. Importantly, the VSC conductivity behavior of
$\{[Si(Pc)O]I_x\}_n$ samples is "metal-like."[62] The results of the
variable temperature conductivity studies also underscore the
robust thermal character of the cofacially arrayed polymers.
$\{[Si(Pc)O]I_x\}_n$ samples could be cycled to 300°C with only minor
deterioration in room temperature conductivity (apparently due to
vaporization of the iodine).

Static magnetic susceptibility measurements by Faraday techi-
niques reveal another characteristic hallmark of "molecular
metals" in the $\{[Si(Pc)O]I_x\}_n$ and $\{[Ge(Pc)O]I_x\}_n$ materials.
Susceptibilities are only weakly paramagnetic (χ_M = 300-500 x
10^{-6} emu after diamagnetic corrections are made) and are only

modestly dependent on temperature down to 77°K[22],[23]. Studies at lower temperature are presently in progress.

THE EFFECTIVENESS OF NONHALOGEN DOPANTS

The availability of a low-dimensional system in which the stacking variable is held constant offers an unusual opportunity, vis-à-vis those factors stabilizing the mixed valent state,[63-66] to investigate the effects of dopants which yield integrated stack structures on reaction with molecular M(Pc) compounds. Thus, it was of interest to learn whether non-halogen oxidants with redox potentials similar to halogens could ever produce a partially oxidized,conductive metallophthalocyanine stack. This would help to define whether the marked efficacy of halogen dopants in forming mixed valent metallomacrocycles reflects, in addition to redox potential, a unique structure-forming role (small yet variable size) or whether other factors (e.g., specific Madelung, exchange, polarization, van der Waals, or core repulsion terms) might be important.

Oxidizing quinones such as those shown below form conductive, mixed valent compounds with a variety of organic donors,[1-5] but all reactions with molecular metallophthalocyanines yield insulators, the structures of which no doubt consist of integrated stacks (Figure 2C).[45],[46] Doping experiments with the $[Si(Pc)O]_n$ polymers were thus carried out by stirring these materials with solutions of the above quinones. The products were characterized by elemental analysis and vibrational spectroscopy. As can be seen in Table 3, large increases in electrical conductivity

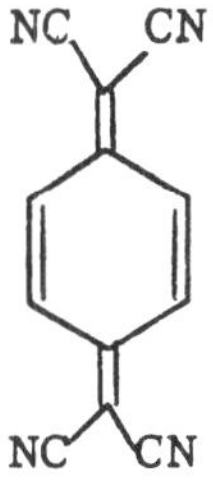

TCNQ 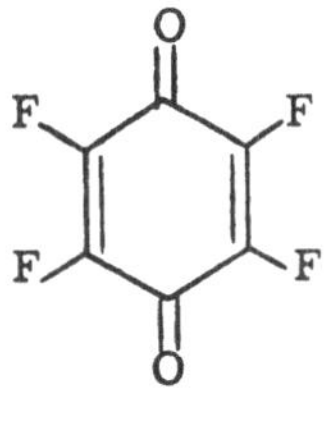   fluoranil 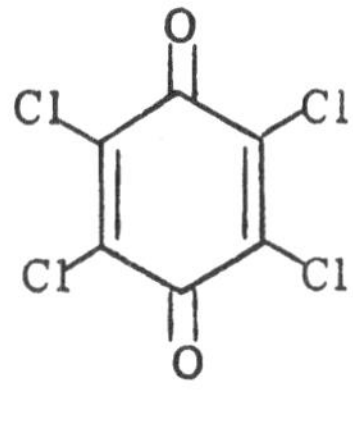   chloranil

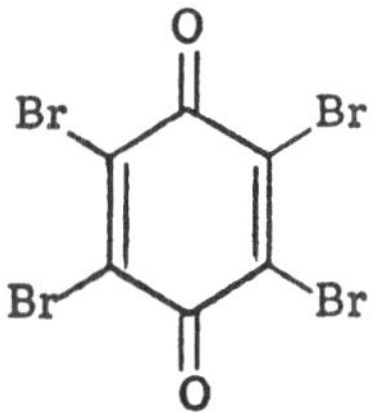

bromanil 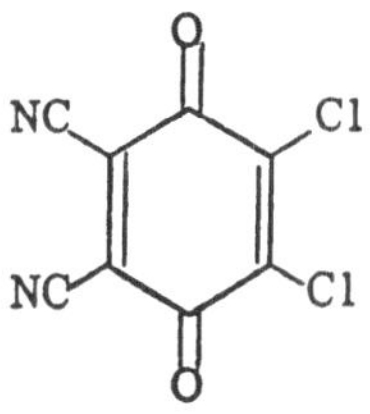   DDQ 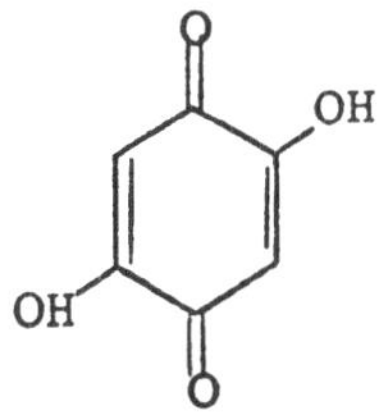   DHB

Table 3. Electrical Conductivity Data for Polycrystalline Samples
of Molecular Phthalocyanines and Cofacial Phthalocyanine
Polymers with Various Dopants

Dopant[a]	Empirical Formula	$\sigma(\Omega^{-1}\ cm^{-1})300°K$	Activation Energy(eV)
none	$[Si(Pc)O]_n$	3×10^{-8}	
I	$\{[Si(Pc)O]I_{1.55}\}_n$	1.4	$0.04 \pm .001$
Br	$\{[Si(Pc)O]Br_{1.00}\}_n$	6×10^{-2}	
K	$\{[Si(Pc)O]K_{1.0}\}_n$	2×10^{-5}	
DDQ	$\{[Si(Pc)O]DDQ_{1.00}\}_n$	2.1×10^{-2}	$0.08 \pm .001$
DDQ	$\{[Si(Pc)O]DDQ_{0.35}\}_n$	6.2×10^{-2}	$0.05 \pm .001$
TCNQ	$\{[Si(Pc)O]TCNQ_{0.50}\}_n$	2.8×10^{-3}	$0.09 \pm .002$
ClA	$\{[Si(Pc)O]ClA_{0.14}\}_n$	1.8×10^{-3}	$0.11 \pm .001$
Flr	$\{[Si(Pc)O]Flr_{0.23}\}_n$	7.2×10^{-4}	$0.13 \pm .001$
Chl	$\{[Si(Pc)O]Chl_{.037}\}_n$	6.9×10^{-4}	$0.13 \pm .002$
Brl	$\{[Si(Pc)O]Brl_{0.84}\}_n$	5.8×10^{-4}	$0.15 \pm .001$
DHB	$\{[Si(Pc)O]DHB_{0.13}\}_n$	3.8×10^{-5}	$0.19 \pm .005$
DDQ	$Ni(Pc)DDQ_{0.11}$	2.5×10^{-7}	0.43 ± 0.004
ClA	$Ni(Pc)ClA_{0.91}$	8.4×10^{-7}	$0.16\raisebox{0.3ex}{\texttildelow}0.002$

a
Flr = fluoranil; Chl = chloranil; Brl = bromanil; DDQ =
dichlorodicyanoquinone; ClA = chloranilic acid; DHB = dihydroxy-
benzoquinone.

accompany the quinone doping. Indeed, the DDQ-doped materials are
as conductive as most of the halogenated polymers.[25,67] The
temperature dependence of the conductivity of some representative
samples is shown in Figure 6. The transport in these materials
is thermally activated and least-squares fits to eq.(2) yield the
activation parameters compiled in Table 3. There are some notable
deviations from a linear ln σ vs. 1/T relationship (e.g., the
$\{[Si(Pc)O]TCNQ_{0.5}\}_n$ sample shown in the Figure) and further
investigations of the reasons for this behavior are in progress.
Perhaps conduction is occurring through <u>both</u> $Pc^{+\delta}$ and $TCNQ^{-\delta}$
stacks with different temperature dependencies. Infrared
spectral studies of the TCNQ-doped materials reveal a displacement
of ν_{CN} to lower frequencies, consistent with electron density uptake by the quinodimethane.[68] Clearly oxidation of the $[Si(Pc)O]_n$
stack by high potential quinones occurs when a segregated stack
structure is enforced. Moreover, this can result in facile charge
transport.

In principle, it should also be possible to partially reduce
phthalocyanines and to create conducting materials by injecting
nonintegral amounts of electron density per site. A number of
attempts have been made in this Laboratory to partially reduce
metallophthalocyanines using alkali metals.[49] In all cases, the
resulting materials were insulators, and it was suspected that
nonstacked materials were being produced. A preliminary experiment was conducted in which $[Si(Pc)O]_n$ was reacted with potassium

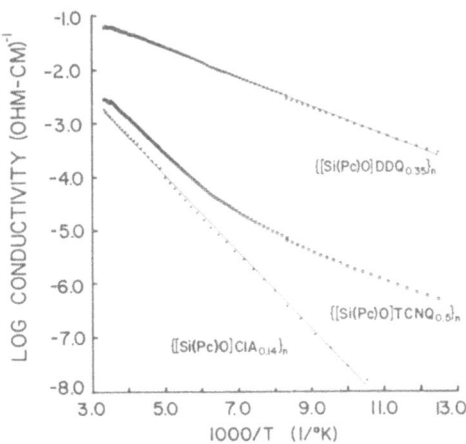

Fig. 6. Variable temperature powder conductivities of the
         siloxane phthalocyaninato cofacial polymer doped with high
         potential quinones. ClA = Chloranilic acid. From ref. 24.

vapor in a sealed tube. The product was collected and handled at all times in an inert atmosphere. As can be seen in Table 3, a significant increase in electrical conductivity accompanies the potassium doping. Further efforts to refine the reductive doping procedure are now in progress.

CONCLUSIONS

Cofacial metallomacrocycle assembly methodology represents what is likely the most powerful approach yet devised for controlling molecular stacking relationships in low dimensional materials. In regard to fundamental understanding, we already have learned a great deal about bandwidth-conductivity and donor-acceptor relationships in conductive materials composed of molecular stacks. However, the surface has only been barely scratched in terms of the exciting research opportunities which await exploitation in this area. Further synthetic work offers the opportunity to make drastic changes in metallomacrocycle identity and electronic structure, stacking distance and bandwidth, interplanar relationships and phonon dynamics, and to correlate these chemical and structural modifications with physical observables. Already, new metal ions[69,70] and bridging functionalities[71,72] have been successfully introduced. Future studies with new types of dopants should provide much important information on donor-acceptor relationships and on those basic factors which stabilize the mixed valent state. A wide range of magnetic, charge transport, and optical experiments remain to be carried out. These experiments should ultimately provide invaluable information on how the chemistry and lattice architecture are connected to some of the fundamental characteristics of the molecular metallic state. Finally, studies of the processing characteristics of the $[M(Pc)O]_n$ materials are just beginning. Already we have learned that it is possible to produce films of $[Si(Pc)O]_n$ and that halogen doping results in substantial increases in electrical conductivity.[72,73] Further efforts in this direction are under way.

ACKNOWLEDGMENTS

This research was generously supported by the Office of Naval Research and by the NSF-MRL program through the Materials Research Center of Northwesern University (grants DMR76-80847A01 and DMR79-23573). TJM is a Camille and Henry Dreyfus Teacher-Scholar. We thank our collaborators Prof. Carl R. Kannewurf, Mr. Joseph W. Lyding, and Dr. Eric A. Mintz for stimulating discussions.

REFERENCES

1. J. T. Devreese, V. E. Evrard, and V. E. Van Doren, eds.,
    "Highly Conducting One-Dimensional Solids," Plenum Press,
    New York (1979).
2. W. E. Hatfield, ed., "Molecular Metals," Plenum Press, New
    York (1979).
3. J. B. Torrance, Accts. Chem. Res. 12:79 (1979).
4. J. S. Miller and A. J. Epstein, eds., "Synthesis and Proper-
    ties of Low-Dimensional Materials," Ann. NY Acad. Sci.,
    313 (1978).
5. H. J. Keller, ed. "Chemistry and Physics of One-Dimensional
    Metals," Plenum Press, New York (1977).
6. J. S. Miller and A. Epstein, J. Prog. Inorg. Chem. 20:1
    (1976).
7. A. Aviram and M. A. Ratner, Chem. Phys. Letters 29:277 (1974)
8. S. Yoshimura and M. Murakami, in reference 4, 269.
9. S. D. Senturia, C. M. Sechen, and J. A. Wishneusky, Appl.
    Phys. Letters 30:106 (1977).
10. C. K. Chiang, S. C. Gau, C. R. Fincher, Jr.; Y. W. Park,
    A. G. MacDiarmid, and A. J. Heeger, Appl. Phys. Letters
    33:18 (1978).
11. M. Ozaki, D. Peebles, B. R. Weinberger, A. J. Heeger, and
    A. G. MacDiarmid, J. Appl. Phys. 51:4252 (1980).
12. A. G. MacDiarmid, private communication.
13. R. S. Potember, T. O. Poehler, A. Rappa, D. O. Cowan, and
    A. N. Bloch, J. Am. Chem. Soc. 102:3659 (1980).
14. Y. Tomkiewicz, E. M. Engler, J. D. Kuptsis, R. G. Schad,
    V. V. Patel, and M. Hatzakis, Extended Abstracts, Electro-
    Chemical Society spring Meeting, St. Louis, May, 1980,
    No. 63.
15. S. Yoshimura, in reference 2, 471, and references therein.
16. C. D. Jaeger and A. J. Bard, J. Am. Chem. Soc. 102:5435
    (1980).
17. E. M. Engler, W. B. Fox, L. V. Interrante, J. S. Miller, F.
    Wudl, S. Yoshimura, A. Heeger, and R. H. Baughman, in ref-
    erence 2, 541.
18. D. Davis, H. Gutfreund, and W. A. Little, Phys. Rev. B 13:
    4766 (1976).
19. J. Bardeen, in reference 1, 373.
20. T. J. Marks and D. W. Kalina in: "Extended Linear Chain Com-
    pounds," J. S. Miller, ed., Plenum Plublishing Corp., NY
    in press.
21. T. J. Marks, Ann. NY Acad. Sci. 313:594 (1978).
22. K. F. Schoch, Jr., B. R. Kundalkar, and T. J. Marks, J. Am.
    Chem. Soc. 101:7071 (1979).

23. T. J. Marks, K. F. Schoch, Jr., and B. R. Kundalkar, Synth. Met. 1:337 (1980).

24. C. W. Dirk, J. W. Lyding, K. F. Schoch, Jr., C. R. Kannewurf, and T. J. Marks, Organic Coatings and Plastics Chem. 43:646 (1980).

25. C. W. Dirk, E. A. Mintz, K. F. Schoch, Jr., and T. J. Marks in: "Organometallic Polymers: Perspectives," C. E. Carraher and J. E. Sheats, eds., Marcel Dekker, NY and J. Macromol. Science-Chemistry, in press.

26. R. C. Teitelbaum, S. L. Ruby, and T. J. Marks, J. Am. Chem. Soc. 102:3322 (1980).

27. R. C. Teitelbaum, S. L. Ruby, and T. J. Marks, J. Am. Chem. Soc. 101:7568 (1979).

28. D. W. Kalina, J. W. Lyding, M. T. Ratajack, C. R. Kannewurf, and T. J. Marks, J. Am. Chem. Soc., in press.

29. M. A. Cowie, A. Gleizes, G. W. Grynkewich, D. W. Kalina, M. S. McClure, R. P. Scaringe, R. C. Teitelbaum, S. L. Ruby, J. A. Ibers, C. R. Kannewurf, and T. J. Marks, J. Am. Chem. Soc. 101:2921 (1979) and references therein.

30. L. D. Brown, D. W. Kalina, M. S. McCLures, S. L. Ruby, S. Schultz, J. A. Ibers, C. R. Kannewurf, and T. J. Marks, J. Am. Chem. Soc. 101:2937 (1979).

31. T. J. Marks, D. F. Webster, S. L. Ruby, and S. Schultz, J. Chem. Soc., Chem. Commun. 444 (1976).

32. A. Gleizes, T. J. Marks, and J. A. Ibers, J. Am. Chem. Soc. 97:3545 (1975).

33. L.-S. Lin, M. S. McClure, J. W. Lyding, M. T. Ratajack, T.-C. Wang, C. R. Kannewurf, and T. J. Marks, J. Chem. Soc., Chem. Commun., in press.

34. M. S. McCLure, L.-S. Lin, T.-C. Whang, M. T. Ratajack, C. R. Kannewurf, and T. J. Marks, Bull. Am. Phys. Soc. 25:315 (1980).

35. J. L. Petersen, C. S. Schramm, D. R. Stojakovic, B. M. Hoffman, and T. J. Marks, J. Am. Chem. Soc. 99:286 (1977).

36. C. S. Schramm, D. R. Stojakovic, B. M. Hoffman, and T. J. Marks, Science 200:47 (1978).

37. R. P. Scaringe, C. J. Schramm, D. R. Stojakovic, B. M. Hoffman, J. A. Ibers, and T. J. Marks, J. Am. Chem. Soc. 102:6702 (1980).

38. T. E. Phillips, R. P. Scaringe, B. M. Hoffman, and J. A. Ibers, J. Am. Chem. Soc. 102:3435 (1980).

39. T. J. Kistenmacher, in reference 4, 333.

40. S. Megtert, J.P. Pougent, and R. Comes, in reference 4, pp. 234.

41. G. D. Stucky, A. J. Schultz, and J. M. Williams, Ann. Rev. Mater. Sci. 7:301 (1977).

42. B. P. Bespalov and V. V. Titov, Russ. Chem. Rev. 44:1091 (1975).

43. D. J. Dahm, P. Horn, G. R. Johnson, M. G. Micles, and J. D. Wilson, J. Cryst. Mol. Struct. 5:27 (1975).
44. F. H. Herbstein, Perspect. Struct. Chem. IV:166 (1971).
45. K. F. Schoch, Jr. and T. J. Marks, unpublished observations.
46. L. Pace and J. A. Ibers, private communication.
47. R. J. Joyner and M. E. Kenney, Inorg. Chem. 82:5790 (1960).
48. J. B. Davison and K. J. Wynne, Macromolecules, 11:186 (1978).
49. K. F. Schoch, Jr. and T. J. Marks, unpublished results at Northwestern University.
50. F. A. Moser and A. L. Thomas, "Phthalocyanine Compounds," Reinhold, New York (1963).
51. F. A. Moser and A. L. Thomas, "Phthalocyanine Compounds," Second Edition, ACS Publications, in press.
52. C. W. Dirk, K. F. Schoch, Jr., and T. J. Marks, manuscript in preparation.
53. G. C. Berry, private communication.
54. D. R. Swift, Ph.D. Thesis, Case Western Reserve University, 1970.
55. C. Glidewell and D. C. Liles, J. Chem. Soc., Chem. Commun. 93 (1979).
56. C. Glidewell and D. C. Liles, Acta Cryst., B34:119 (1978).
57. C. Glidewell and D. C. Liles, Acta Cryst., B34:124 (1978).
58. C. Glidewell and D. C. Liles, Acta Cryst., B34:129 (1978).
59. R. C. Teitelbaum, Ph.D. Thesis, Northwestern University, August 1979.
60. K. Seeger, "Semiconductor Physics," Springer-Verlag, NY, 1973, 483.
61. L. B. Coleman, Rev. Sci. Instrum. 49:48 (1978).
62. K. F. Schoch, Jr., J. W. Lyding, C. R. Kannewurf, and T. J. Marks, manuscript in preparation.
63. A. J. Epstein, N. O. Lipari, D. J. Sandman, and P. Nielsen, Phys. Rev., B 13:1569 (1976).
64. R. M. Metzger, In reference 1d, 145.
65. J. B. Torrance and B. D. Silverman, Phys. Rev., B, 15:788 (1977).
66. R. M. Metzger and A. N. Bloch, J. Chem. Phys., 63:5098 (1975).
67. K. F. Schoch, Jr. and T. J. Marks, unpublished results; Bull. Am. Phys. Soc. 25:315 (1980).
68. R. P. Van Duyne, M. R. Suchanski, J. M. Lakovits, A. R. Siedle, K. D. Parks, and T. M. Cotton, J. Am. Chem. Soc. 101:2832 (1979).
69. P. M. Kuznesof, K. J. Wynne, R. S. Nohr, and M. E. Kenney, J. Chem. Soc., Chem. Commun. 121 (1980).
70. R. S. Nohr, K. J. Wynne, and M. E. Kenney, Polymer Preprints, in press. We thank these authors for a preprint.
71. C. W. Dirk and T. J. Marks, Bull. Am. Phys. Soc. 25:315 (1980).

72. C. W. Dirk, J. W. Lyding, K. F. Schoch, Jr., C. R. Kannewurf,
     and T. J. Marks, Abstracts, <u>Fall Meeting of the American
     Chemical Society</u>, Las Vegas, August 1980.
73. C. W. Dirk, M. S. McClure, K. F. Schoch, Jr., C. R.
     Kannewurf, and T. J. Marks, unpublished results at
     Northwestern University.

UTILIZATION OF POLYACETYLENE, $(CH)_x$, IN THE FABRICATION OF RECHARGEABLE BATTERIES

Paul J. Nigrey, David MacInnes, Jr., David P. Nairns,
Alan G. MacDiarmid, and Alan J. Heeger*

Department of Chemistry and Department of Physics*
University of Pennsylvania, Philadelphia, Pa. 19104

INTRODUCTION

Polyacetylene, $(CH)_x$, is the first example of a covalent organic polymer which may be chemically doped either p- or n-type to
give a series of semiconductors and ultimately "organic metals".[1]
The electrical conductivity can be varied over twelve orders of
magnitude depending on the dopant concentration.[1] Detailed studies
have shown that a semiconductor-metal transition occurs at dopant
concentrations near 1 mole percent. In the metallic range, (1-10
mole percent) the conductivity increases at a relatively slow rate
up to a value of $\sim 10^3$ ohm^{-1}cm^{-1} at room temperature. We have shown
previously that ~ 0.1 mm thick films of $(CH)_x$ may also be controllably
doped electrochemically in a very simple, rapid procedure, either
in aqueous or non-aqueous solution, through the semiconducting to
the metallic regime.[2] Thus, the use of aqueous KI solutions or
CH_2Cl_2 solutions of, e.g., $[(n-C_4H_9)_4N]^+[ClO_4]^-$, $[(n-C_4H_9)_4N]^+[AsF_6]^-$,
etc. yield flexible, golden-silvery films of $[CHI_{0.07}]_x$ ($\sigma_{25°C}=$
9.7 ohm^{-1}cm^{-1}), $[CH(ClO_4)_{0.0645}]_x$ ($\sigma_{25°C}=970$ ohm^{-1}cm^{-1}), and
$[CH(AsF_4)_{0.077}]_x$ ($\sigma_{25°C}=553$ ohm^{-1}cm^{-1}), respectively. In the doped
state, the polyacetylene is believed to exist as the stabilized
polycarbonium ion, $(CH^{+y})_x$, with the corresponding number of mono-
valent counter anions such that the overall composition is
$[(CH^{+y})A_y^-]_x$.[1] In the case of iodine doping, it has been shown that
at least a large amount of the iodine exists as the I_3^- ion.[3]

When electrochemical doping is carried out at an applied potential near 9 V using ca. 0.1 mm thick cis-rich $(CH)_x$ film as the
anode, the metallic-doped state is generally reached in 30-60
minutes.[2] This relatively rapid doping is undoubtedly assisted by
the fact that the film consists of an interwoven network of ca.

200 Å $(CH)_x$ fibrils which fill only about one third of the volume of the film.[1] The surface area of the $(CH)_x$ fibrils in the film is 40-60 meters^2/gram.[1] A 1 cm^2 piece of film 0.1 mm in thickness therefore has an effective surface area of approximately 2.5×10^3 cm^2.

The anode reaction occurring upon doping, using $[(n-C_4H_9)_4N]^+A^-$ as an electrolyte, involves oxidation of $(CH)_x$, viz.,

$$(CH)_x \rightarrow (CH^{+y})_x + xye^-. \tag{1}$$

The corresponding cathode reaction, at an inert electrode such as platinum, is

$$xy[(n-C_4H_9)_4N]^+ + xye^- \rightarrow xy''[(n-C_4H_9)_4N]^{o''}. \tag{2}$$

The $[(n-C_4H_9)_4N]^+$ upon discharge will break down in a complex manner to give amines and olefins, etc. The overall electrochemical doping reaction may therefore be summarized as

$$(CH)_x + xy[(n-C_4H_9)_4N]^+[A]^- \rightarrow [(CH^{+y})A_y^-]_x + xy''[(n-C_4H_9)_4N]^{o''}.$$

$$\tag{3}$$

It will be noted that the A^- ion does not enter into the electro-chemical reaction. It serves only as a negative counter ion to preserve electrical neutrality in the system.

An important extension of the electrochemical doping study is the attempted electrochemical "undoping" of $[(CH^{+y})A_y^-]_x$ to reform the original $(CH)_x$. This also has now been accomplished. It is clearly apparent that the doping reaction given by equation (1), involving partial oxidation of the $(CH)_x$, may also be regarded as a possible battery "charging reaction" to produce a doped polymer which could be utilized as a cathode-active battery material. The discharge reactions are the reverse of those given above.

A number of different dopant ions, solvents, electrolytes, electrolyte concentrations and battery configurations have been investigated, only one of which is described below in order to illustrate the potential application of $(CH)_x$ in batteries. This system is not necessarily the most preferred one nor are its para-meters necessarily optimized to give the best output characteristics.

EXPERIMENTAL

Materials and Reagents

Cis-rich $(CH)_x$ film (ca. 80% cis isomer; ca. 0.1 mm thickness; density ca. 0.4 gm/cc) was prepared as previously described.[4,5]

Propylene carbonate (Aldrich Reagent grade) was stirred over lithium chips for 14 days. This was used to make a 0.3 M solution of LiClO$_4$ which was passed through Woelm activated basic alumina to yield a colorless solution. Anhydrous LiClO$_4$ (Alfa-Ventron) was dried by melting in vacuo before use. Lithium metal (Alfa-Ventron Co.) was scraped under a LiClO$_4$ solution in propylene carbonate with a knife immediately before use.

Doping Experiments

Doping experiments were carried out in a dry, inert atmosphere using a glass vessel containing the LiClO$_4$/propylene carbonate solution. A strip of (CH)$_x$ film (3 cm x 1.5 cm x 0.01 cm), to the top of which a platinum wire had been attached by mechanical pressed contact, was placed in the solution so that the film and wire were completely immersed. A platinum counter electrode (3 cm^2) was placed in the solution at a distance of ~12 cm from the (CH)$_x$ working electrode. A Ag/Ag$^+$(0.1 M AgNO$_3$ in CH$_3$CN) reference electrode was then placed in the container ~1 cm from the (CH)$_x$ working electrode. The three electrodes were attached to a Princeton Applied Research (PAR) Model 173 potentiostat/galvanostat equipped with a PAR Model 179 digital coulometer.

Electrochemical doping of the (CH)$_x$ with (ClO$_4$)$^-$ was accomplished by holding it at a potential of 1.0 V with respect to the Ag/Ag$^+$ reference electrode. After 10 seconds, the film was withdrawn so that the top of the platinum wire was ~3 mm above the surface of the solution. As doping proceeded and the conductivity of the film increased, the current also increased. The extent of doping was followed by observing the value of the current. At the very beginning, the current was 40 µA and after 33 minutes it had increased to 4 mA. It stayed constant at this value for an additional four minutes and then began to decrease. Doping was discontinued at this stage, the film was washed several times with propylene carbonate, then with CH$_2$Cl$_2$ and then dried in vacuo with constant pumping overnight. The flexible, golden film had a 4-probe conductivity of ~700 ohm^{-1}cm^{-1} at room temperature. Its composition from the number of coulombs passed was [CH(ClO$_4$)$_{0.043}$]$_x$. This composition is not necessarily expected to agree exactly with that obtained from elemental analysis since there is uncertainty in the extent of doping of that 3 mm portion of the film above the solution. A portion of that part of the film which had been completely immersed in the solution was analyzed commercially.[6] Its composition was [CH(ClO$_4$)$_{0.058}$]$_x$. Found: C=63.23%; H=5.71%; Cl=10.21%; O (by difference)=20.85%. Calc. for CHCl$_{0.058}$O$_{0.232}$: C=63.96%; H=5.37%; Cl=10.95%; O=19.77%.

Battery Experiments

A strip of (CH)$_x$ film (0.5 cm x 1 cm x 0.01 cm), either

undoped or previously doped, was placed in the propylene carbonate
solution as described previously. Platinum wire was attached to the
top of the film as in the doping experiments. The lithium counter
electrode (0.5 cm x 2 cm x 0.1 cm) was placed ~0.5 cm from the
(working) $(CH)_x$ electrode. A similar strip of lithium metal was
placed approximately equidistantly from the working and counter
electrodes and was used as the reference electrode.

In some experiments, $(CH)_x$ film previously doped with perchlorate
was used initially. This was because the composition of that 3 mm
portion of the film above the surface of the solution would then be
the same as that below the surface. When undoped $(CH)_x$ film was used
initially, it could not necessarily be assumed that the composition
of that portion of the film above the surface of the solution was
fully doped or whether it might become more doped on each charging
cycle during a series of charge/discharge cycles. When undoped $(CH)_x$
film was employed, it was first electrochemically doped in situ at
a constant current of 1 mA for 26 minutes to give a film of composition $[CH(ClO_4)_{0.06}]_x$ (6% doping) as calculated from the weight of
the film employed and the number of coulombs passed. In some
instances, film doped only to 5% was employed.

Discharge/charge studies were carried out using two different
methods: (i) constant current investigations employed a model 173
galvanostat where the change in voltage versus a lithium reference
electrode during a discharge or charge cycle was monitored by means
of a Keithley model 177 digital multimeter connected to a Houston
Omnigraphics 2000 XY recorder. Voltage versus time measurements
were recorded at a number of different discharge currents (0.1-4 mA);
(ii) short circuit studies involved discharge through a Keithley
model 177 digital multimeter, where current and voltage were measured
until the cell was completely discharged. The battery was recharged
to its original level after each discharge cycle by means of a d.c.
power supply set at 4 V or by a constant current method using the
PAR 173 galvanostat at 1 mA.

RESULTS AND DISCUSSION

Doping of $(CH)_x$ with $(ClO_4)^-$

We have shown previously that $(CH)_x$ film could be doped readily
to composition $[CH(ClO_4)_{0.0645}]_x$ from a 0.5 M solution of
$[(n-C_4H_9)_4N]^+[ClO_4]^-$ in CH_2Cl_2 at an applied potential of 9V.[2] In
the present study, we have shown that doping may also be carried
out conveniently using a 0.3 M solution of $LiClO_4$ in propylene carbonate at an applied potential of 1.0 V (versus Ag/Ag^+) during 35
minutes at room temperature. When $LiClO_4$ is used instead of
$[(n-C_4H_9)_4N]^+A^-$, and Li metal is used instead of Pt at the cathode,
then the reaction corresponding to equation 2 is

$$xyLi^+ + xye^- \rightarrow xyLi^o \tag{4}$$

The overall reaction corresponding to equation 3 is then

$$(CH)_x + xyLi^+(ClO_4)^- \rightarrow [(CH^{+y})(ClO_4)_y^-]_x + xyLi^o . \tag{5}$$

Battery Experiments

The simplest battery configuration uses a piece of free-standing (CH)$_x$ film, nearly all of which is immersed in a propylene carbonate solution of LiClO$_4$. The top of the film is attached by means of a wire to the positive terminal of a 9V dry cell or d.c. power supply. The negative terminal is attached to a lithium metal electrode immersed in the solution. The battery is charged, i.e., the film is doped to composition corresponding approximately to [CH(ClO$_4$)$_{0.06}$]$_x$ in 30 minutes.

The open circuit voltage of a 6% predoped film, i.e., [CH(ClO$_4$)$_{0.06}$]$_x$ was ~3.7 V and the short circuit current was approximately 25 mA for an 0.5 cm^2 (~3 mg) piece of film. This varied from ~15 mA to 35 mA depending on the sample of (CH)$_x$ film employed and the extent of doping, etc. The high current density may be related to the large effective surface area of the (CH)$_x$ fibrils in the film. Because of the small mass of the film used in these experiments, the short circuit current falls rapidly as the battery discharges. For example, after ~30 seconds of short circuit discharge, the voltage was essentially unchanged and I$_{sc}$ was ~4 mA. More than half of the total coulombs evolved were given off during this period. After 1 minute the voltage was ~3.3 V and I$_{sc}$ was ~2 mA. After a total of ~3.5 minutes, the voltage began to fall rapidly and I$_{sc}$ decreased to ~0.3 mA. At this stage, the dopant concentration was ~3%, a concentration close to that defining the metal-semiconductor transition where the conductivity of p-doped (CH)$_x$ films begins to fall rapidly. These V$_{oc}$ and I$_{sc}$ values give an energy density of ~80 watt hours/lb. based on the weights of the combined cathode and anode active materials consumed in the reverse of the chemical reaction given by equation 5. Successive short circuit discharging to ~3% dopant concentration (V$_{oc}$=2.7 V; I$_{sc}$=1.0 mA) and recharging to ~6% dopant concentration showed that the number of coulombs involved in the charging and discharging process were identical. Results of experiments carried out under more rigorously controlled conditions are given below.

A number of experiments were carried out at constant charge and discharge currents, with the charge and discharge currents having the same value. The change in voltage was recorded throughout each experiment. A typical experiment involved three complete charge/discharge cycles using currents of 0.550 mA and a 0.5 cm^2 piece of cis-rich (CH)$_x$ film (total surface area of both sides, 1.0 cm^2).

There is no particular significance in the choice of this current
value; it is simply of a convenient magnitude. The charging voltage
at the end of a charge cycle was 4.1 V. This fell instantly to 3.7 V
at the very beginning of a discharge cycle. At the end of a discharge
cycle it was 3.2 V. Each charge and discharge cycle involved 120
millicoulombs and took four minutes for the charge cycle and four
minutes for the discharge cycle.[5]

The composition at the beginning of each of the charge cycles
at 0.550 mA was $[CH(ClO_4)_{0.045}]_x$, and at the end of each charge
cycle it was $[CH(ClO_4)_{0.050}]_x$. The difference in composition (involving 0.005 mole of $(ClO_4)^-$) is obtained from the coulombs passed.
The number of coulombs used in the charge cycle is exactly equal,
within experimental error, to the number of coulombs released in
the discharge cycle. To the accuracy of the measurements, the $(CH)_x$
electrode was found to be reversible with no observable degradation
after a number of charge/discharge cycles. For example, after
twenty charge and twenty discharge cycles at constant currents
ranging from 0.1 to 1 mA, a given $(CH)_x$ electrode showed no change
in its charging or discharging characteristics. It should also be
noted that the battery does not spontaneously lose its charge. Thus
a charged battery after standing for 48 hours still showed the same
initial voltage (3.7 V) and discharged at 0.550 mA exactly as before.

Longer discharge periods at lower controlled constant current
conditions also gave good energy densities. For example, a single,
constant current discharge at 0.550 mA resulted in a drop in voltage
to 2.7 V after 26 minutes and a 2 mA discharge gave a voltage of
2.7 V after 7 minutes. Voltage recovery was noted under certain
conditions when a period of discharge was followed by a rest period
during which no current was drawn from the cell. For example, after
the last discharge cycle in the experiment described above, where
three complete charge/discharge cycles at 0.550 mA had been carried
out, a V_{oc} of 3.2 V was observed. The cell was then permitted to
stand for ~3 minutes. It then again displayed a voltage of 3.7 V
and a further discharge cycle at 0.550 mA had similar characteristics
to those found previously.

Although the energy density values given above are interestingly large, they are not necessarily the maximum obtainable. Preliminary experiments suggest that it may be possible to dope $(CH)_x$
to even higher levels with perchlorate; furthermore, other dopants
are known where doping up to at least a value of 10% has been
obtained. In a real battery operation, the weight of the cathode
current collector electrode must also be added to the weight of the
cathode-active material. In the $(CH)_x$ cathode configuration,
because the $(CH)_x$ is a free-standing metal itself, it serves both
as the current collector and as the cathode-active material, thus
saving significantly in total weight. Furthermore, as mentioned
earlier, the (metallic) conductivity of $[CH(ClO_4)_y]_x$ remains essen-

tially constant for values of y≈0.01 to 0.06. Hence, <u>at least</u> 5/6 of the material can be gainfully used electrochemically without significant change of those current collecting properties related to its conductivity.

The results of the present studies on the use of doped polyacetylene as the cathode-active material in batteries are most encouraging. They suggest the possibility of relatively inexpensive lightweight rechargeable batteries made with polyacetylene which may have a variety of potential technological applications.

ACKNOWLEDGMENT

The authors wish to thank Dr. G. C. Farrington for helpful discussions and the Office of Naval Research for support of this research.

REFERENCES

1. A. G. MacDiarmid and A. J. Heeger, Organic Metals and Semiconductors: The Chemistry of Polyacetylene, <u>Synth</u>. <u>Met</u>., 1:101 (1979/80); A. J. Heeger and A. G. MacDiarmid, Organic Metals and Semiconductors: The Chemistry of Polyacetylene, (CH)$_x$, and Its Derivatives, <u>in</u>: "The Physics and Chemistry of Low Dimensional Solids," L. Alcácer, ed., D. Reidel Publishing Co., Dordrecht, Holland (1979).

2. P. J. Nigrey, A. G. MacDiarmid and A. J. Heeger, Electrochemistry of Polyacetylene, (CH)$_x$: Electrochemical Doping of (CH)$_x$ Films to the Metallic State, <u>J.C.S</u>. <u>Chem</u>. <u>Comm</u>., 594 (1979).

3. S. L. Hsu, A. J. Signorelli, G. P. Pez and R. H. Baughman, Highly Conducting Iodine Derivatives of Polyacetylene: Raman, XPS and X-ray Diffraction Studies, <u>J</u>. <u>Chem</u>. <u>Phys</u>., 69:106 (1978).

4. H. Shirakawa and S. Ikeda, Preparation and Morphology of As-Prepared and Highly Stretch-Aligned Polyacetylene, <u>Synth</u>. <u>Met</u>., 1:175 (1979/80).

5. Similar results were obtained with (CH)$_x$ film kindly supplied by Rohm and Haas Co., Bristol, Pa.

6. Galbraith Laboratories, Inc., Knoxville, Tn.

INDEX